T0140803

Christian Nolde

Global Regularity and Uniqueness of Solutions in a Surface Growth Model Using Rigorous A-Posteriori Methods

λογος

Augsburger Schriften zur Mathematik, Physik und Informatik
Band 32

Edited by:
Professor Dr. B. Schmidt
Professor Dr. B. Aulbach
Professor Dr. F. Pukelsheim
Professor Dr. W. Reif
Professor Dr. D. Vollhardt

Bibliographic information published by the Deutsche Nationalbibliothek

The Deutsche Nationalbibliothek lists this publication in the Deutsche Nationalbibliografie; detailed bibliographic data are available in the Internet at http://dnb.d-nb.de .

ISBN 978-3-8325-4453-9
ISSN 1611-4256

Logos Verlag Berlin GmbH
Comeniushof, Gubener Str. 47,
10243 Berlin
Tel.: +49 030 42 85 10 90
Fax: +49 030 42 85 10 92
INTERNET: http://www.logos-verlag.de

Global Regularity and Uniqueness of Solutions in a Surface Growth Model Using Rigorous A-Posteriori Methods

Dissertation

zur Erlangung des akademischen Grades

Dr. rer. nat.

eingereicht an der

Mathematisch-Naturwissenschaftlich-Technischen Fakultät
der

UNIVERSITÄT AUGSBURG

von

Christian Nolde

Augsburg, Dezember 2016

Universität Augsburg
Mathematisch-Naturwissenschaftlich-
Technische Fakultät

Gutachter:

- Prof. Dr. Dirk Blömker, Universität Augsburg

- Prof. Dr. Malte Peter, Universität Augsburg

Datum der mündlichen Prüfung: 17.02.2017

Contents

CHAPTER 1

Introduction

The aim of this thesis is to set up rigorous a-posteriori methods to prove the existence and uniqueness of a smooth global solution to a fourth order semi-linear parabolic partial differential equation (PDE). More specific, we consider the following surface growth equation for the height $u(t, x) \in \mathbb{R}$ at time $t > 0$ over a point $x \in [0, 2\pi]$

$$u_t = -u_{xxxx} - (u_x{}^2)_{xx} \qquad x \in [0, 2\pi], \; t \in [0, T] \qquad (1.1)$$

with periodic boundary conditions and subject to a moving frame, which yields the zero-average condition $\int_0^{2\pi} u(x, t) \, \mathrm{d}x = 0$.

This equation, usually with additional noise term, was introduced as a phenomenological model for the growth of vapor deposited amorphous surfaces [SP94]; [RLH00], and was also used to describe ion-sputtering processes, where a surface is eroded by an ion-beam [CVG05]. The one dimensional equation appears as a model for the boundaries of terraces in the epitaxy of silicon [FV06]. A more detailed list of references can be found in [BR12] and [BR15].

On the mathematical side, the existence of global weak solutions and local unique and smooth solutions is shown in [BR09] and expanded in [BR12]. Unfortunately, due to a gap in the regularity requirements for the initial condition, these weak solutions are neither unique nor smooth. It is also known, that the complex-valued surface growth equation in fact does experience a blowup in finite time for certain initial values, where the mass is transfered faster to higher Fourier modes than it dissipates [BR15]. It remains if this is also true for the real-valued function.

1

An exhaustive summary of the equation's derivation and known mathematical properties, in the stochastic as well as the deterministic case, is given by Blömker and Romito in [BR15].

The famous sibling

The, by far, most famous problem for existence, uniqueness and regularity of solutions to a semi-linear parabolic partial differential equation is the Navier-Stokes Problem [Fef06]. This problem poses the question, if the motion of a fluid is uniquely determined and smooth for all time, as long as the initial data is smooth. Sometimes, it is also described as the problem whether the water inside your bathtub might explode (which is a nice pun to the mathematical "blowup" we are looking for). In the year 2000, the Clay Mathematics Institute published their famous list of seven Millennium Prize Problems including the Navier-Stokes Problem. Despite all this attention, the problem remains unsolved until today. For more information about the Navier-Stokes Problem and why it is so hard to solve, there is a detailed blog post by Terrence Tao [07] where he analyzes, why methods, that work for other problems, fail when it comes to Navier-Stokes.

Computer aided proofs

An increasingly popular method, especially for problems which are not (yet) solvable in the classical way, are computer assisted proofs. These range from "simple" brute force methods, as used in the proof for the famous four color problem (see [AH89]), to extremely involved artificial intelligence programs which use automated reasoning to proof or even deduce new theorems (e.g. [HKM16]).

A very common type, and the one we will refer to in this thesis, is rigorous numerics, where quantities calculated by the computer are set-valued rather than a floating point number. These sets represent the error bound for the true value and take into account machine precision, round-off and truncation of values and arithmetic operations.

Unsurprisingly, the fact that computers are involved in mathematical proofs was controversial in the beginning (and for some people maybe still is), as these are usually not verifiable in the way proofs used to be, and that they could suffer from bugs (errors) in software (the program itself, compiler/interpreter, operating system) or hardware (memory, processor). The answer to this is, that computer assisted proofs have to be verified differently. In some sense they are like rigorous experiments, which have to be replicated using different hardware and software to avoid bugs. There is also the concept of "formal verification" to prove the correctness of algorithms which is useful here. For the rigorous numerics, it might be the best to look at them like a "really huge" calculation inside a proof. This is not uncommon to mathematical proofs, just that in this case the calculations are done by a computer rather than a human being and they just have to be verified accordingly.

The Fields Medallist William P. Thurston wrote over 20 years ago in [Thu94] about that topic:

> The rapid advance of computers has helped dramatize this point, because computers and people are very different. For instance, when Appel and Haken completed a proof of the 4-color map theorem using a massive automatic computation, it evoked much controversy. I interpret the controversy as having little to do with doubt people had as to the veracity of the theorem or the correctness of the proof. Rather, it reflected a continuing desire for human understanding of a proof, in addition to knowledge that the theorem is true.

> On a more everyday level, it is common for people first starting to grapple with computers to make large-scale computations of things they might have done on a smaller scale by hand. They might print out a table of the first 10,000 primes, only to find that their printout isn't something they really wanted after all. They discover by this kind of experience that what they really want is usually not some collection of "answers" – what they want is understanding.

Nowadays, rigorous methods for proving numerically the existence of solutions for PDEs are an active research field. In addition to the approach we will take, there are other methods based on topological arguments like the Conley index, see [Mai+08]; [DLM07]; [Zgl10], for example. For solutions of elliptic PDEs there are also methods using Brouwer's fixed-point theorem, as discussed in the review article [Plu08] and the references therein, or for a small summary for fixed-point methods in rigorous numerics see [BL15].

Inspiration for our approach

In [Che+07]; [DR08]; [RMS13]; [RS08] Chernyshenko, Constantin, Dashti, Marín-Rubio, Robinson, Sadowski and Titi applied the idea of rigorous numerics to the Navier-Stokes Problem. They set up the analytic framework to show the existence of a smooth global solution using rigorous numerics, but did not carry out the actual numeric implementation. A similar approach using an integral equation based on the mild formulation was proposed in [MP08]; [MP11]. The inspiration to establish similar methods for the surface growth equation is the work of Robinson et al. [Che+07], but a similar method was proposed by Morosi and Pizzocchero in [MP12].

The key ingredients for the analytic framework are a) a global regularity result for small initial data and b) the existence of local unique and smooth solutions. For Navier-Stokes a) is provided by Leray in [Ler34] and b) by Koch and Tataru in [KT01]. The good news is, that both properties hold in a similar way for the surface growth equation, which was shown by Blömker and Romito in [BR09].

Now, the main idea of the method is the derivation of a scalar ODE for the H^1-norm of the difference between an arbitrary approximation, which satisfies the boundary conditions, and the solution. The coefficients of this ODE depend only on the numerical data (or any other approximation used). As long as the solution of the ODE stays finite, one can rely on the continuation property of unique local solutions, and thus have a smooth unique solution up to a blowup time of the ODE. This is improved by the smallness property and its corollary, a time property which proves global existence if this solution does not experience a blowup until a given time T^*, or if the solution is falling below a certain threshold. Please note, that this method is only able to verify the existence of a smooth and unique global solution. It is unable to verify a blowup.

As we limit ourselves to the scalar surface growth equation, the numerical implementation should be a lot easier compared to 3D Navier-Stokes. Moreover, for the two-dimensional case the situation of energy estimates seems even worse, as global existence could only be established in H^{-1} using the non-standard energy $\int_0^{2\pi} e^{u(x)} \, \mathrm{d}x$, see [Win11] for details. Nevertheless, we believe that it should be possible to treat the 2D case using similar methods, but the analysis becomes more delicate since in two dimensions H^1 is the critical space (see [BR09]; [BR12]).

1.1 Outline

In Chapter 2, we will set up all necessary analytic results for the surface growth equation. We will carry out the energy estimate to obtain the differential inequality (2.1) governing the error between solution and approximation. After that, we will formulate the smallness and time conditions

(Theorem 2.2 and Theorem 2.3) which allow us to prove global existence in finite time, what is very important for practical purpose. Following, we propose two analytical methods: one, based on the standard Gronwall Lemma, enforces a 'small data' hypothesis and adds little to standard analytical existence proofs. The second is based on an explicit analytical upper bound to the ODE solution (Lemma 2.5 and Corollary 2.7).

Chapter 3 will now apply these results to the differential inequality (2.1) (Theorem 3.1 and Theorem 3.2). We will also test Corollary 2.7 with restarting, which means that one applies the analytical upper bound on a succession of small intervals of length $h > 0$ to the numerical solution and then restarts the argument (Theorem 3.3). We will also give a formal calculation which indicates that the upper bound from the third method in the limit of stepsize to zero converges to the solution of the ODE. To finish the chapter, we will make first (non rigorous) numerical experiments to compare the bounds delivered by the methods. The simulations show that all methods work (i.e. they are able to reach at least one of our two goals, smallness and time), but Method 3 (Corollary 2.7 with restarting, Theorem 3.3) is superior as it is not as sensitive to crucial quantities like $\|\varphi_{xx}\|_\infty$ and the residual as its competitors. Please note, that the results from Chapters 2 and 3 have already been published as joint work with Dirk Blömker and James C. Robinson in [BNR15].

In Chapter 4, we will limit the influence of $\|\varphi_{xx}\|_\infty$ by improving our initial energy estimate (2.1). Therefore we analyze the spectrum of the linearized operator (here $Lv = -v_{xxxx} + (v_x\varphi_x)_{xx}$, where φ is some given numerical data) with a rigorous numerical method, which in the case of an unstable linear operator yields substantially better results, at the price of a significantly higher computational time (Theorem 4.1). This approach is similar to [NH09]; [NKK12]; [Liu15] where this was proposed in a slightly different context. An important aim for our estimate is, that it has to be efficiently computable. Therefore, not every theoretical result might be suitable for us. When we apply this new estimate to our ODE (2.1), we also free all constants from Young inequalities, to further improve the estimate by solving an optimization problem for given times, and obtain the new ODE bound.

In Chapter 5, we will apply Method 3 (Theorem 3.3) to our old (2.1) and new bound with eigenvalue estimate (4.5). Further, we will do all calculations and estimates to rigorously compute all quantities needed by our methods. Therefore, we will define φ as the linear interpolation between the numerically obtained grid points. This leads to the rigorous computable methods that we will denote by Method 4 (Theorem 5.1) without eigenvalue estimate, and Method 5 (Theorem 5.2) with eigenvalue estimate.

Chapter 6 will now deliver the rigorous (except interval arithmetic) simulations of Methods 4 and 5. The results show that both methods work, but Method 5 with the eigenvalue estimate is a huge improvement over Method 4. Initial values like $\sin(x)$ which are already far away from the scope of analytic results are successfully handled by both methods. They even have plenty of room for an additional error,which means we can be optimistic that interval arithmetic should not change the outcome. As the methods are still vulnerable to initial values where $\|\varphi_{xx}\|_\infty$ is large (in this case large means e.g. $u_0 = \sin(3x)$), the problem is not solved for all initial values.

1.2 General Setting and Notation

As solutions to our surface growth equation (1.1) are subject to periodic boundary conditions on $[0, 2\pi]$ with mean average, we are working on the space

$$\mathcal{H} = \left\{ u : \mathbb{R} \to \mathbb{R} \ : \ 2\pi\text{-periodic}, \ \int_0^{2\pi} u(x) \ \mathrm{d}x = 0 \right\}$$

with scalar product $\langle \cdot, \cdot \rangle$ and norm

$$\|u\| = \left(\int_0^{2\pi} |u(x)|^2 \ \mathrm{d}x \right)^{1/2}.$$

We further define the Sobolev-spaces

$$\mathcal{H}^k = \{ u \in \mathcal{H} \ : \ \partial_x^k u \in L^2([0, 2\pi]) \}.$$

Note, that by periodicity $u \in \mathcal{H}^1$ implies $u_x \in \mathcal{H}$. Moreover, we have Poincare-inequality with optimal constant 1

$$\|u\| \leq \|u_x\| \quad \text{for all } u \in \mathcal{H}^1$$

and thus $\|u_x\|$ is a norm on $\mathcal{H}^1$, equivalent to the standard H^1-Sobolev norm. Furthermore, interpolation inequality (Theorem A.6)

$$\|u_x\|^2 \leq \|u_{xx}\| \|u\| \quad \text{for all } u \in \mathcal{H}^2$$

and Agmon's inequality (Theorem A.5)

$$\|u\|_\infty^2 \leq \|u\| \|u_x\|$$

also hold with constant 1.
    We will use the abbreviations

$$u_0 := u(x, 0)$$

for the initial value of solutions or approximations to our surface growth model,

$$\|f\|_{-1} := \|f\|_{\mathcal{H}^{-1}}$$

for the $\mathcal{H}^{-1}$-norm and

$$\|f\|_{\infty} := \|f\|_{L^{\infty}([0,2\pi])}$$

for the $L^{\infty}([0,2\pi])$-norm.

1.3   Acknowledgments

Firstly, I would like to express my sincere gratitude to my advisor Prof. Dirk Blömker for the continuous support, guidance and advice he gave me during my studies and the patience he had when necessary. Also, having the possibility to discuss issues at any time is something that can not be taken for granted and I am extremely thankful for. I really enjoyed and appreciated every discussion we had, be it during coffee break or late at night on skype.

Further, I want to thank Prof. James C. Robinson for the invitation to Warwick to discuss parts of this thesis.

I would also like to thank every member of the chair "Nichtlineare Analysis" for the nice time we had during the last years. It was a pleasure to meet so many different people from all over the world. Our social activities, discussions, joint teaching duties and seminars—I enjoyed every part of it.

Finally, I want to thank my family and friends for their support during the last years.

CHAPTER 2

A-priori Analysis

The aim of this chapter is to establish upper bounds for the $\mathcal{H}^1$-norm of the error

$$d(x,t) := u(x,t) - \varphi(x,t),$$

where u is a solution to our surface growth equation (1.1) and φ is any arbitrary, but sufficiently smooth approximation, that satisfies the boundary conditions. Since we know φ, if we can control the $\mathcal{H}^1$-norm of d then we control the $\mathcal{H}^1$-norm of u.

A very important property of the surface growth equation (1.1), and basically the foundation of this work, is the existence of local solutions, which are smooth in space and time. Their existence is given by the following theorem from [BR09] (Theorem 3.1) (There is actually a much simpler proof, that is sufficient for our setting, in [Hen81] using sectorial operators and Banach's fixed point theorem)

Theorem 2.1. Let $u_0 \in H^1$, then there exists a time $\tau(u_0) > 0$ such that there is a unique solution $u \in C^0([0, \tau(u_0)), H^1)$ satisfying

  1) if $\tau(u_0) < \infty$, then $\limsup\limits_{t \to \tau(u_0)} \|u(t)\|_{H^1} = \infty$.

  2) u is C^∞ in both, space and time, for all $(t, x) \in (0, \tau(u_0)) \times [0, 2\pi]$.

Note, that the theorem implies that lack of blowup in H^1 is sufficient to ensure that the solution exists for all time and is smooth. In particular, all of the manipulations we make in what follows are valid until the blowup time.

9

From now on, we consider the solutions with initial data in $\mathcal{H}^1$ whose existence is guaranteed by Theorem 2.1, and approximations $\varphi \in \mathcal{H}^4$ in space and H^1 in time.

2.1 Energy Estimate

In this section we prove the key estimate (2.1) on which the theorems of the following sections are based.

As mentioned before, let u be a smooth solution with initial data in $\mathcal{H}^1$ whose existence is guaranteed by Theorem 2.1, and approximations $\varphi \in \mathcal{H}^4$ in space and H^1 in time.

If we use the surface growth equation (1.1) to find the evolution of $d(x, t)$ and define the residual of the approximation φ by

$$\mathrm{Res} := \varphi_t + \varphi_{xxxx} + (\varphi_x{}^2)_{xx},$$

then we have

$$d_t = -d_{xxxx} - (u_x{}^2)_{xx} + (\varphi_x{}^2)_{xx} - \mathrm{Res}.$$

By replacing u with $d + \varphi$ we obtain

$$d_t = -d_{xxxx} - (d_x{}^2)_{xx} - 2(d_x\varphi_x)_{xx} - \mathrm{Res}.$$

As d is sufficiently smooth for $t > 0$, the $\mathcal{H}^1$-norm holds that

$$\frac{1}{2}\partial_t\|d_x\|^2 = \langle \partial_t d_x, d_x \rangle = -\langle d_{xx}, d_t \rangle$$
$$= \underbrace{\langle d_{xx}, d_{xxxx} \rangle}_{\mathrm{A}} + \underbrace{2\langle d_{xx}, (d_x\varphi_x)_{xx} \rangle}_{\mathrm{B}} + \underbrace{\langle d_{xx}, (d_x{}^2)_{xx} \rangle}_{\mathrm{C}} + \underbrace{\langle d_{xx}, \mathrm{Res} \rangle}_{\mathrm{D}},$$

where $\langle \cdot, \cdot \rangle$ is the L^2 scalar product.

Note that when using integration by parts, due to $d \in \mathcal{H}^4$ we know that $d \in C^3_{\mathrm{per}}$ and thus, as we have at most three derivatives in the boundary terms, these cancel out.

Now consider these terms separately. Integrating by parts we obtain

$$\mathrm{A} = -\|d_{xxx}\|^2$$

(compare Theorem A.9).

Secondly,

$$\mathrm{B} = -2\int_0^{2\pi} d_{xxx}(d_x\varphi_x)_x \, \mathrm{d}x$$

$$= -2 \int_0^{2\pi} d_{xxx} d_{xx} \varphi_x \; \mathrm{d}x - 2 \int_0^{2\pi} d_{xxx} d_x \varphi_{xx} \; \mathrm{d}x$$
$$= - \int_0^{2\pi} (d_{xx}{}^2)_x \varphi_x \; \mathrm{d}x - 2 \int_0^{2\pi} d_{xxx} d_x \varphi_{xx} \; \mathrm{d}x$$
$$= \int_0^{2\pi} (d_{xx})^2 \varphi_{xx} \; \mathrm{d}x - 2 \int_0^{2\pi} d_{xxx} d_x \varphi_{xx} \; \mathrm{d}x$$

and so

$$|\mathrm{B}| \leq \|d_{xx}\|^2 \|\varphi_{xx}\|_\infty + 2\|d_{xxx}\|\|d_x\|\|\varphi_{xx}\|_\infty$$
$$\leq 3\|d_{xxx}\|\|d_x\|\|\varphi_{xx}\|_\infty$$
$$\leq \frac{1}{4}\|d_{xxx}\|^2 + 9\|d_x\|^2 \|\varphi_{xx}\|_\infty^2,$$

using interpolation (Theorem A.6) and Young's inequality (Theorem A.2). For C we have

$$\mathrm{C} = - \int_0^{2\pi} (d_x{}^2)_x d_{xxx} \; \mathrm{d}x = -2 \int_0^{2\pi} d_x d_{xx} d_{xxx} \; \mathrm{d}x,$$

hence using Agmon's inequality (Theorem A.5), interpolation, and Young's inequality,

$$|\mathrm{C}| \leq 2\|d_x\|\|d_{xx}\|_\infty \|d_{xxx}\|$$
$$\leq 2\|d_x\|\|d_{xx}\|^{\frac{1}{2}} \|d_{xxx}\|^{\frac{3}{2}}$$
$$\leq 2\|d_x\|^{\frac{5}{4}} \|d_{xxx}\|^{\frac{7}{4}}$$
$$\leq \frac{1}{4}\|d_{xxx}\|^2 + \frac{7^7}{4}\|d_x\|^{10},$$

and for the remaining term

$$|\mathrm{D}| \leq \| \operatorname{Res} \|_{-1}\|d_{xxx}\| \leq \frac{1}{4}\|d_{xxx}\|^2 + \| \operatorname{Res} \|_{-1}^2.$$

Combining these estimates and applying Poincaré inequality (Theorem A.3) with the optimal constant $\omega = 1$, we obtain

$$\frac{1}{2}\partial_t \|d_x\|^2 \leq -\frac{1}{4}\|d_{xxx}\|^2 + \frac{7^7}{4}\|d_x\|^{10} + \| \operatorname{Res} \|_{-1}^2 + 9\|d_x\|^2 \|\varphi_{xx}\|_\infty^2$$
$$\leq \frac{7^7}{4}\|d_x\|^{10} + \left(9\|\varphi_{xx}\|_\infty^2 - \frac{1}{4}\right)\|d_x\|^2 + \| \operatorname{Res} \|_{-1}^2.$$

Thus

$$\partial_t \|d_x\|^2 \leq \frac{7^7}{2} \|d_x\|^{10} + \left(18\|\varphi_{xx}\|_\infty^2 - \frac{1}{2} \right) \|d_x\|^2 + 2\| \operatorname{Res} \|_{-1}^2, \qquad (2.1)$$

which is a scalar differential inequality of type

$$\dot{\xi} \leq b\xi^5 + (a(t) - c)\xi + f(t), \qquad (2.2)$$

and by standard ODE comparison principles (Theorem A.8) a solution of the equality in (2.1) provides an upper bound for $\|d_x\|^2$.

2.2 Time and Smallness Conditions

We need two important properties of the surface growth model, which we will prove now. These are for equations like Navier–Stokes well known facts, namely: that smallness of the solution implies global uniqueness and that solutions are actually small after some time by energy-type estimates. These results go back to Leray ([Ler34]), more modern discussions can be found in [CF88] (Theorem 9.3) and in a setting that parallels the treatment here in [RS08]. For our model similar results for the critical $H^{1/2}$-norm can be found in [BR09]. But for our numerical evaluations, we need to derive the precise values of constants in the $\mathcal{H}^1$-norm, which were not determined before.

First, if the $\mathcal{H}^1$-norm of a solution u at any time is smaller than some constant ε_0, we have regularity after that time. If the $\mathcal{H}^1$-norm is also bounded up to this time, we have global regularity of u.

Theorem 2.2 (Smallness Condition). *If for some $t \in [0,T]$ one has that $\|u_x\|$ is finite on $[0,t]$ and*

$$\|u_x(t)\| < \frac{1}{2} =: \varepsilon_0,$$

then we have global regularity (and thus uniqueness) of the solution u on $[0,\infty)$.

Proof. This is established by almost the same estimates derived for the parts (A) and (C) in Section 2.1 and Young's inequality with constant $\delta > 0$. To be more precise:

$$\frac{1}{2}\partial_t \|u_x\|^2 = -\|u_{xxx}\|^2 + \int_0^{2\pi} u_{xx}(u_x{}^2)_{xx} \, \mathrm{d}x$$
$$\leq -\|u_{xxx}\|^2 + 2\|u_{xxx}\|^{\frac{7}{4}} \|u_x\|^{\frac{5}{4}}$$

$$\leq -\|u_{xxx}\|^2 + 2 \cdot \left(\delta\|u_{xxx}\|^2 + \left(\frac{8}{7}\delta\right)^{-7} \cdot \frac{1}{8}\|u_x\|^{10} \right)$$

$$\leq -\|u_{xxx}\|^2 \left(1 - 2\delta - \left(\frac{8}{7}\delta\right)^{-7} \cdot \frac{1}{4}\|u_x\|^8 \right).$$

If the term in parentheses is positive, i.e.

$$0 < 1 - 2\delta - \left(\frac{8}{7}\delta\right)^{-7} \cdot \frac{1}{4}\|u_x\|^8$$

$$\Rightarrow \quad \|u_x\|^8 < (1 - 2\delta) \cdot 4 \cdot \left(\frac{8}{7}\delta\right)^7,$$

then we obtain a global bound on $\|u_x\|^2$. The optimal choice for the constant from Young inequality is $\delta = \frac{7}{16}$ as this maximizes the right hand side of the inequality. With this value it follows, that if $\|u_x(t)\| < \frac{1}{2}$ we have a negative derivative and the norm decays over time and is therefore bounded. $\qquad \square$

The second property is that, based on the smallness condition, we can determine a time T^*, only depending on the initial value u_0, such that $\|u_x(T^*)\| < \varepsilon_0$.

Theorem 2.3 (Time Condition). *If a solution u is regular up to time*

$$T^*(u_0) := \frac{1}{2\varepsilon_0^2}\|u_0\|^2 = 2\|u_0\|^2,$$

then we have global regularity of the solution u.

At the risk of laboring the point, we only need to verify regularity of a solution starting at u_0 up to time $T^*(u_0)$, and from that point on regularity is automatic.

Proof. As an a-priori estimate we have (see Theorem A.9)

$$\frac{1}{2}\partial_t\|u\|^2 = -\|u_{xx}\|^2$$

and thus

$$\int_0^T \|u_x(s)\|^2 \, \mathrm{d}s \leq \int_0^T \|u_{xx}(s)\|^2 \, \mathrm{d}s \leq \frac{1}{2}\|u_0\|^2,$$

where we used the Poincaré inequality (Theorem A.3) with constant $\omega = 1$. If we now assume that $\|u_x(s)\| > \varepsilon_0$ for all $s \in [0, T]$, then

$$T\varepsilon_0^2 < \frac{1}{2}\|u_0\|^2 \quad \text{or} \quad T < \frac{1}{2\varepsilon_0^2}\|u_0\|^2.$$

This means, that if we stay bounded until time $T^* := \frac{1}{2\varepsilon_0^2}\|u_0\|^2$, we know that $\|u_x(t)\| \leq \varepsilon_0$ for at least one $t \in [0, T^*]$ and we have global regularity by the smallness condition. $\qquad \square$

2.3 ODE Estimates

We will now show several methods to bound solutions of ODEs of the type (2.2). In this section we give the results for the scalar ODE, and present applications in the next section.

Let us first state a lemma of Gronwall type, based on comparison principles for ODEs, for which we will only give the idea of a proof.

Lemma 2.4 (Gronwall). *Let $a, b \in L^1([0,T], \mathbb{R})$ and $x \in W^{1,1}([0,T], \mathbb{R}) \cap C^0([0,T], \mathbb{R})$ such that*

$$\dot{x} \leq a(t)x + b(t) \qquad \forall t \in [0,T].$$

Then for all $t \in [0,T]$

$$x(t) \leq \exp\left(\int_0^t a(s) \, \mathrm{d}s\right) x(0) + \int_0^t \exp\left(\int_s^t a(r) \, \mathrm{d}r\right) b(s) \, \mathrm{d}s \, .$$

Idea of Proof. Consider the function

$$u(t) = x(t) \exp\{-\int_0^t a(s)ds\} \quad \text{with} \quad u'(t) \leq b(t) \exp\{-\int_0^t a(s)ds\}.$$

Integrating and solving for x yields the result.   □

Lemma 2.5. *Consider two functions $x, u \in W^{1,1}([0,T], \mathbb{R}_0^+) \cap C^0([0,T], \mathbb{R}_0^+)$ such that*

$$\dot{x} \leq c(t)x^p + e(t) \qquad x(0) = x_0$$

with $p > 1$, $c \in L^1([0,T], \mathbb{R}_0^+)$ and $e \in L^1([0,T], \mathbb{R}_0^+)$, and let u be the solution of

$$\dot{u} = c(t)u^p \qquad u(0) = x_0 + \int_0^T e(s) \, \mathrm{d}s.$$

Then $x(t) \leq u(t)$ for all $t \in [0,T]$.

Proof. First note, that if $e \equiv 0$ on $[0,T]$ then by using the standard comparison principle it follows that $u(t) \geq x(t)$ for all $t \in [0,T]$.

So now we assume that $\int_0^T e(s) \, \mathrm{d}s > 0$. For a contradiction, suppose that there exists a time $t^* \in [0,T]$ such that $t^* := \inf\{t > 0 : x(t) = u(t)\}$. Because of the continuity of $u(t)$ and $x(t)$, and $u(0) > x(0)$ due to our initial assumption $\int_0^T e(s) \, \mathrm{d}s > 0$, it follows that $t^* > 0$. From the definition $u(t) > x(t)$ for all $t \in [0, t^*)$, and thus

$$0 = u(t^*) - x(t^*) \geq u(0) - x(0) - \int_0^{t^*} e(s) \, \mathrm{d}s + \int_0^{t^*} c(s)(u(s)^p - x(s)^p) \, \mathrm{d}s$$

$$= \int_{t^*}^{T} e(s) \, \mathrm{d}s + \int_{0}^{t^*} c(s)(u(s)^p - x(s)^p) \, \mathrm{d}s,$$

which is strictly positive provided that $\int_{0}^{t^*} c(s) \, \mathrm{d}s > 0$.

If $\int_{0}^{t^*} c(s) \, \mathrm{d}s = 0$, then as $c \geq 0$ we obtain

$$x(t) \leq x(0) + \int_{0}^{t} e(s) \, \mathrm{d}s \leq x(0) + \int_{0}^{T} e(s) \, \mathrm{d}s = u(t) \quad \forall t \in [0, t^*],$$

and we can repeat the above argument on the interval $[t^*, T]$ to obtain a contradiction. $\qquad\square$

Theorem 2.6. *Assume* $x \in W^{1,1}([0,T], \mathbb{R}_0^+) \cap C^0([0,T], \mathbb{R}_0^+)$ *such that*

$$\dot{x} \leq c(t)x^p + e(t), \qquad x(0) = x_0$$

with $p > 1$, $c \in L^1([0,T], \mathbb{R}_0^+)$ *and* $e \in L^1([0,T], \mathbb{R}_0^+)$. *Then for all* $t \in [0,T]$, *as long as the right-hand side is finite,*

$$x(t) \leq \left(x_0 + \int_0^t e(s) \, \mathrm{d}s \right) \left\{ 1 - (p-1)\left[x_0 + \int_0^t e(s) \, \mathrm{d}s \right]^{p-1} \int_0^t c(s) \, \mathrm{d}s \right\}^{-\frac{1}{p-1}}.$$

Proof. Given the setting of Lemma 2.5, we can solve for $u(t)$. As $\mathrm{d}u = c(t)u^p \, \mathrm{d}t$, a straightforward calculation shows that

$$u(t) = u(0)\left(1 - (p-1)u(0)^{p-1} \int_0^t c(s) \, \mathrm{d}s \right)^{-\frac{1}{p-1}}$$

as long as the right-hand side is finite. Thus for all $t \in [0,T]$, as long as the right-hand side is finite,

$$x(t) \leq \left(x_0 + \int_0^T e(s) \, \mathrm{d}s \right)$$
$$\times \left\{ 1 - (p-1)\left[x_0 + \int_0^T e(s) \, \mathrm{d}s \right]^{p-1} \int_0^t c(s) \, \mathrm{d}s \right\}^{-\frac{1}{p-1}}$$

This holds particularly when $T = t$. $\qquad\square$

We now extend this result to differential inequalities of the form

$$\dot{x} \leq b(t)x^p + a(t)x + f(t),$$

where $p > 1$, $f, b \in L^1([0,T], \mathbb{R}_0^+)$ and $a \in L^1([0,T], \mathbb{R})$, as our inequality (2.1) is of this type.

Corollary 2.7. *Assume* $x \in W^{1,1}([0,T], \mathbb{R}_0^+) \cap C^0([0,T], \mathbb{R}_0^+)$ *such that*

$$\dot{x} \le b(t)x^p + a(t)x + f(t),$$

with $p > 1$, *b*, $f \in L^1([0,T], \mathbb{R}_0^+)$ *and* $a \in L^1([0,T], \mathbb{R})$. *Then for all* $t \in [0,T]$, *as long as the right-hand side is finite,*

$$x(t) \le e^{A(t)} \left(x_0 + \int_0^t \tilde{f}(s) \, ds \right)$$

$$\times \left\{ 1 - (p-1) \cdot \left[x_0 + \int_0^t \tilde{f}(s) \, ds \right]^{p-1} \int_0^t \tilde{b}(s) \, ds \right\}^{-\frac{1}{p-1}}$$

where

$$\tilde{b}(t) = b(t)e^{(p-1)A(t)}, \quad \tilde{f}(t) = e^{-A(t)}f(t), \quad and \quad A(t) = \int_0^t a(s) \, ds.$$

Proof. Consider the substitution $y(t) = e^{-A(t)}x(t)$ with $A(t) = \int_0^t a(s) \, ds$. It follows that

$$
\begin{aligned}
\dot{y} &= -a(t)y + e^{-A(t)}\dot{x} \\
&\le -a(t)y + e^{-A(t)}(b(t)x^p + a(t)x + f(t)) \\
&= \underbrace{b(t)e^{(p-1)A(t)}}_{\tilde{b}(t)} \, y^p + \underbrace{e^{-A(t)}f(t)}_{\tilde{f}(t)}
\end{aligned}
$$

with $\tilde{b}(t) \ge 0$ and $\tilde{f}(t) \ge 0$ for all $t \in [0,T]$. Here we can apply Theorem 2.6 and obtain

$$y(t) \le \left(y_0 + \int_0^t \tilde{f} \, ds \right) \left\{ 1 - (p-1) \cdot \left[y_0 + \int_0^t \tilde{f} \, ds \right]^{p-1} \int_0^t \tilde{b} \, ds \right\}^{-\frac{1}{p-1}}.$$

Now substitute back with $x(t) = e^{A(t)}y(t)$. $\square$

CHAPTER 3

Verification Methods

In this chapter, we outline three techniques for numerical verification. All of them are based on the key estimate (2.1) for the difference d between an arbitrary smooth approximation φ and a smooth local solution. The first method is additionally based on the simple Gronwall Lemma 2.4, the second on Corollary 2.7, and the third is similar to the second method, but restarts the estimation after a series of short time-steps.

3.1 First Method

By assuming a bound for $\|d_x(t)\|$, we can reduce (2.1) to a linear ODE which has the right type for the simple Gronwall Lemma 2.4. With this lemma, we can establish an, at least initially, better error estimate. The drawback however is, that if our estimate exceeds the previously set bound, we can not make assertions on $\|d_x(t)\|$ any longer.

Theorem 3.1 (Method 1). *Let* $K^* = (2 \cdot 7^7)^{-1/8}$. *As long as*

$$\|d_x(0)\|^2 e^{A(t)} + 2 \int_0^t \|\operatorname{Res}(s)\|_{-1}^2 e^{(A(t)-A(s))} \, ds \leq K^*, \qquad (3.1)$$

we have

$$\|d_x(t)\|^2 \leq \|d_x(0)\|^2 e^{A(t)} + 2 \int_0^t \|\operatorname{Res}(s)\|_{-1}^2 e^{(A(t)-A(s))} \, ds,$$

where $A(t) = -\frac{1}{4}t + 18 \int_0^t \|\varphi_{xx}(\tau)\|_\infty^2 \, d\tau$.

Note, that the condition in (3.1) involves only the numerical approximation φ.

Proof. It follows from the inequality (2.1) that as long as $\|d_x\|^8 \leq (2 \cdot 7^7)^{-1}$ we obtain

$$\partial_t \|d_x\|^2 \leq -\frac{1}{4} \|d_x\|^2 + 2\| \operatorname{Res} \|_{-1}^2 + 18\|d_x\|^2 \|\varphi_{xx}\|_\infty^2.$$

Now we can apply Lemma 2.4 to deduce that

$$\|d_x(t)\|^2 \leq \|d_x(0)\|^2 \exp \left\{ -\frac{t}{4} + 18 \int_0^t \|\varphi_{xx}(\tau)\|_\infty^2 \, \mathrm{d}\tau \right\}$$
$$+ 2 \int_0^t \| \operatorname{Res}(s) \|_{-1}^2 \exp \left\{ -\frac{t-s}{4} + 18 \int_s^t \|\varphi_{xx}(\tau)\|_\infty^2 \, \mathrm{d}\tau \right\} \mathrm{d}s.$$

$\square$

Again, please note that if the bound from this exceeds K^*, Theorem 3.1 makes no assertions on $\|d_x\|^2$.

3.2   Second Method

This is the more sophisticated method based on direct application of Corollary 2.7.

Theorem 3.2 (Method 2). *As long as the right-hand side is finite, the following inequality holds for $d(t)$:*

$$\|d_x(t)\|^2 \leq e^{A(t)} \left(\|d_x(0)\|^2 + \int_0^t \tilde{f}(s) \, \mathrm{d}s \right)$$
$$\times \left\{ 1 - 4 \left[\|d_x(0)\|^2 + \int_0^t \tilde{f}(s) \, \mathrm{d}s \right]^4 \int_0^t \tilde{b}(s) \, \mathrm{d}s \right\}^{-1/4}$$

with

$$\tilde{b}(t) = \frac{7^7}{2} \, e^{4A(t)}, \qquad \tilde{f}(t) = 2e^{-A(t)} \| \operatorname{Res}(t) \|_{-1}^2$$

and

$$A(t) = -\frac{t}{2} + \int_0^t 18\|\varphi_{xx}(s)\|_\infty^2 \, \mathrm{d}s.$$

Again, the condition for regularity provided by the theorem depends only on the numerical solution φ.

Proof. Apply Corollary 2.7 (CP-Type II) to our inequality (2.1). The corresponding functions are

$$b(t) = \frac{7^7}{2}, \quad a(t) = 18\|\varphi_{xx}(t)\|_\infty^2 - \frac{1}{2}, \quad f(t) = 2\| \operatorname{Res}(t) \|_{-1}^2,$$

which immediately give us the statement of the theorem. $\square$

3.3 Third Method (Second Method with Restarting)

The previous method can be further improved by introducing something that can be best described as "restarting". Instead of estimating over the whole time interval $[0, T]$ at once, we estimate to some smaller t^* and use the resulting upper bound as the new initial value.

Theorem 3.3 (Method 3). *Given any arbitrary partition $\{t_i\}_{0 \le i \le n}$ of the interval $[0, T]$ with $t_0 = 0$ and $t_n = T$, then by Theorem 3.2 we have for all $1 \le i \le n$*

$$z(0) := \|d_x(0)\|^2$$

$$\|d_x(t_i)\|^2 \le \mathrm{e}^{A(t_i)} \left(z(t_{i-1}) + \int_{t_{i-1}}^{t_i} \tilde{f}(s) \, \mathrm{d}s \right)$$

$$\times \left\{ 1 - 4 \left[z(t_{i-1}) + \int_{t_{i-1}}^{t_i} \tilde{f}(s) \, \mathrm{d}s \right]^4 \int_{t_{i-1}}^{t_i} \tilde{b}(s) \, \mathrm{d}s \right\}^{-1/4}$$

$$=: z(t_i)$$

as long as the right-hand side is finite, where for $t \in (t_{i-1}, t_i]$

$$\tilde{b}(t) = \frac{7^7}{2} \, \mathrm{e}^{4A(t)}, \qquad \tilde{f}(t) = 2\mathrm{e}^{-A(t)} \| \operatorname{Res}(t) \|_{-1}^2$$

and

$$A(t) = -\frac{1}{2}(t - t_{i-1}) + \int_{t_{i-1}}^{t} 18\|\varphi_{xx}(s)\|_\infty^2 \, \mathrm{d}s.$$

Proof. Given some arbitrary partition $\{t_i\}_{0 \le i \le n}$ of the interval $[0, T]$ with $t_0 = 0$ and $t_n = T$, we define our new method as follows.

First, we apply Theorem 3.2 to the interval $[0, t_1]$. Here we start with

$$z(0) := \|d_x(0)\|^2$$

and thus obtain on the right-hand side of the interval

$$\|d_x(t_1)\|^2 \le \mathrm{e}^{A(t_1)} \left(z(0) + \int_0^{t_1} \tilde{f}(s) \, \mathrm{d}s \right)$$

$$\times \left\{ 1 - 4 \left[z(0) + \int_0^{t_1} \tilde{f}(s) \, \mathrm{d}s \right]^4 \int_0^{t_1} \tilde{b}(s) \, \mathrm{d}s \right\}^{-1/4}$$

$$=: z(t_1).$$

This defines the upper bound for $\|d_x(t_1)\|^2$ as $z(t_1)$. In the next step, $z(t_1)$ is taken as the new "initial value" when we apply Theorem 3.2 to the interval $[t_1, t_2]$.

$$\|d_x(t_2)\|^2 \leq e^{A(t_2)}\left(z(t_1) + \int_{t_1}^{t_2} \tilde{f}(s)\,\mathrm{d}s\right)$$

$$\times \left\{(1 - 4\left[z(t_1) + \int_{t_1}^{t_2} \tilde{f}(s)\,\mathrm{d}s\right]^4 \int_{t_1}^{t_2} \tilde{b}(s)\,\mathrm{d}s\right\}^{-1/4}$$

where $\tilde{b}(t), \tilde{f}(t)$ are defined as before, only $A(t)$ for $t \in (t_{i-1}, t_i]$ changes to

$$A(t) = -\frac{1}{2}(t - t_{i-1}) + \int_{t_{i-1}}^{t} 18\|\varphi_{xx}(s)\|_\infty^2\,\mathrm{d}s.$$

This procedure is now repeated for every interval of the partition.   □

Lastly, we want to give an informal argument that this method converges to a solution of the ODE as $h \to 0$. Only informal, because a rigorous proof would be extensive for adding little to the understanding.

Conjecture 3.4. Method 3 as given by Theorem 3.3, where the step-size h is given by $\max_{1\leq i<n}(t_i - t_{i-1}) = h$, converges for $h \to 0$ to a solution of the ODE (2.1) bounding $\|d_x\|$.

Sketch of proof. Let $z(t)$ be a smooth interpolation of the discrete points $z(t_i)$, $i = 1, 2, \ldots$ and $h = t_{j+1} - t_j$. Then

$$\partial_t z(t_j) = \frac{z(t_{j+1}) - z(t_j)}{h} + \mathcal{O}(h)$$

Using $\int_{t_j}^{t_{j+1}} g\,\mathrm{d}s = g(t_j)h + \mathcal{O}(h^2)$ and the abbreviations $z(t_j) = z_j$, $A_j = A(t_j)$ and $\mathrm{Res}_j = \|\,\mathrm{Res}(t_j)\|_{-1}^2$, we obtain from Theorem 3.3

$$\partial_t z(t_j) = \frac{1}{h}\left[\frac{e^{A_{j+1}}(z_j + h\tilde{f}_j + \mathcal{O}(h^2))}{(1 - 4[z_j + h\tilde{f}_j + \mathcal{O}(h^2)]^4 h\tilde{b}_j + \mathcal{O}(h^2))^{1/4}} - z_j\right] + \mathcal{O}(h)$$

$$\overset{\text{Taylor}}{=} \frac{1}{h}\left[\frac{e^{A_{j+1}}(z_j + h\tilde{f}_j)}{(1 - 4[z_j + h\tilde{f}_j]^4 h\tilde{b}_j)^{1/4}} - z_j\right] + \mathcal{O}(h)$$

Using $\tilde{b}_j = \frac{7^7}{2}e^{4A_j} = \frac{7^7}{2}$ and $\tilde{f}_j = 2e^{-A_j}\,\mathrm{Res}_j = 2\,\mathrm{Res}_j$, as $A_j = 0$ yields

$$\partial_t z(t_j) = \frac{1}{h}\left[\frac{e^{A_{j+1}}(z_j + 2h\,\mathrm{Res}_j)}{(1 - [z_j + 2h\,\mathrm{Res}_j]^4 \cdot 2h7^7)^{1/4}} - z_j\right] + \mathcal{O}(h)$$

$$= \frac{1}{h}\left[\frac{e^{A_{j+1}}z_j + 2he^{A_{j+1}}\,\mathrm{Res}_j - z_j\sqrt[4]{1 - 7^7 2h[z_j + 2h\,\mathrm{Res}_j]^4}}{\sqrt[4]{1 - 7^7 2h[z_j + 2h\,\mathrm{Res}_j]^4}}\right] + \mathcal{O}(h)$$

$$= \frac{1}{h}\left[e^{A_{j+1}}z_j + 2he^{A_{j+1}}\,\mathrm{Res}_j - z_j\sqrt[4]{1 - 7^7 2h[z_j + 2h\,\mathrm{Res}_j]^4}\right] + \mathcal{O}(h).$$

Now using $\sqrt[4]{1-x} = 1 - \frac{1}{4}x + O(x^2)$ (like before) and $A_{j+1} = O(h)$ leads to

$$\partial_t z(t_j) = 2\mathrm{e}^{A_{j+1}} \operatorname{Res}_j + \frac{1}{h}(\mathrm{e}^{A_{j+1}} - 1)z_j + z_j \frac{1}{2} 7^7 [z_j + 2h\operatorname{Res}_j]^4 + O(h)$$

$$= 2\operatorname{Res}_j + 2A'(t_j)z_j + \frac{1}{2}7^7 z_j^5 + O(h).$$

Recall that $\operatorname{Res}_j = \|\operatorname{Res}(t_j)\|_{-1}^2$ and $A'(t_j) = -\frac{1}{2} + 18\|\varphi_{xx}(t_j)\|_\infty^2$, and we recover that z solves (2.1) with equality in the limit $h \to 0$. □

3.4   Numerical Comparison

In this section, we will compare the three previously established methods numerically. Note, that the simulations are not rigorous yet. We do not implement interval arithmetic which would be necessary to take inevitable rounding errors into account and further, we do not compute a strict upper bound of the methods. For the latter see Chapter 5. However, as our aim is to illustrate the general behavior and feasibility of the three methods, this should be sufficient at this point. Moreover, adding interval arithmetic is just a technical straight forward task, where we expect that the additional error does not affect the picture that much if used with adaptive step-size.

Although our methods allow φ to be any arbitrary approximation, that satisfies the boundary conditions, it should be a reasonable choice, i.e. close to an expected solution, for the methods to be successful.

For our simulations, we calculate an approximate solution using a spectral Galerkin scheme with N Fourier modes in space and a semi-implicit Euler scheme with step-size h in time, yielding the values $\varphi(t)$ for $t = 0, h, 2h, \dots$. Further details about the numerical implementation can be found in Appendix B. To calculate the residual of φ, these values are interpolated piecewise linear in time.

There are two ways to show global regularity using the numerical methods of the previous section:

- show that the solution exists until the time $T^*(u_0)$ (from Theorem 2.3), since the solution is regular after this time; or

- show that $\|\varphi_x(t)\| + \|d_x(t)\| < \varepsilon_0$ for some $t > 0$, since then Theorem 2.2 guarantees global regularity.

In all of our figures the maximum time is T^*, as defined by Theorem 2.3, rounded to the first decimal digit $+0.1$ as a small padding, as long as no

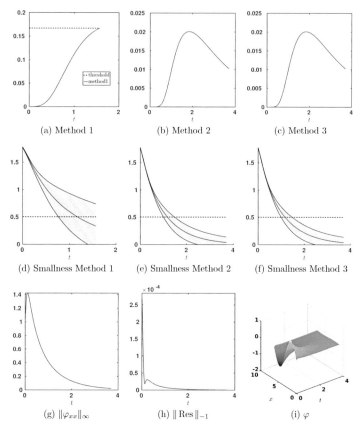

Figure 3.1: Initial value $u_0 = \sin(x)$, $N = 128$ Fourier modes and step-size $h = 10^{-5}$. Methods 2 and 3 show global existence of a smooth solution as they stay bounded until time T^* and even fulfill the smallness criterion before time T^*. Method 1 fails as it hits its threshold at approximately $t = 1.5 < T^*$ and also before the smallness criterion is reached.

blowup occurred or threshold was hit. The values for T^* can be found in Table 3.1.

In Figure 3.1 we have an initial value of $u_0 = \sin(x)$, $N = 128$ Fourier modes and a step-size of $h = 10^{-5}$. As Methods 2 and 3 stay bounded up to time T^*, we would have shown global existence of a unique smooth solution in a fully rigorous simulation. Method 1 fails because it hits its threshold at approximately $t = 1.57$, which is smaller than T^*. In the "Smallness plots" (Figures 3.1d, 3.1e and 3.1f) the gray area is the area around $\|\varphi_x(t)\|$ (the solid blue line in the middle) with distance $\|d_x(t)\|$, which is estimated by the respective method. The red dashed line indicates the critical value $\varepsilon_0 = \frac{1}{2}$ of the smallness criterion (Theorem 2.2). Note that $\|u_x(0)\| = \sqrt{\pi} \gg \frac{1}{2}$ which implies that we do not have any analytic results for this initial value. In order to fulfill the smallness criterion, there has to be a time τ where the upper border of the gray area is below the red dashed line.

To sum up, in this example we would have been able to show global regularity for this initial value with Methods 2 and 3 by smallness and time criteria, whereas both fail for Method 1. In Figure 3.2 we achieve global existence also with Method 1 by decreasing the step-size to $h = 10^{-6}$. All other parameters stay unchanged.

Figures 3.3, 3.4 and 3.5 have minimal alterations of the initial value with $u_0 = \sin(x) + 0.25 \sin(2x)$, $u_0 = \sin(x) + 0.225 \sin(2x)$ and $u_0 = \sin(x) + 0.2 \sin(2x)$ respectively. We can see that these small alterations are sufficient to change our results completely. In Figure 3.3 all three methods fail, in Figure 3.4 only Method 1 fails and in Figure 3.5 all three methods succeed.

Note, that the bounds for all methods decline approximately at $t \approx 2$ because φ is almost identical 0. In Figure 3.4, this decline is still too late for Method 1 as it has already hit its threshold. A further small reduction of the initial value in Figure 3.5 is now enough to keep also Method 1 alive. Also $\|u_x(0)\| \gg \frac{1}{2}$ for these three initial values again.

Figure 3.3 is also interesting, as the numerical approximation alone does get below the smallness threshold, but not if you take into account our error bound (what is not too surprising if you remember that all approximations tend to zero).

Table 3.1 is a collection of the actual values for the Figures 3.1 to 3.5. It is worth to note, that, at least for the current dataset, if we have global existence by one criterion, then also by the other. The smallness criterion is also reached significantly earlier than the time criterion.

As the bounds from Methods 2 and 3 are in all of our simulations almost identical, we use an artificial example to illustrate the difference between these methods. In Figure 3.6 we do this by artificially setting a constant, relatively large $\| \operatorname{Res} \|_{-1}^2 = 0.5$ and an also constant, but smaller second deriva-

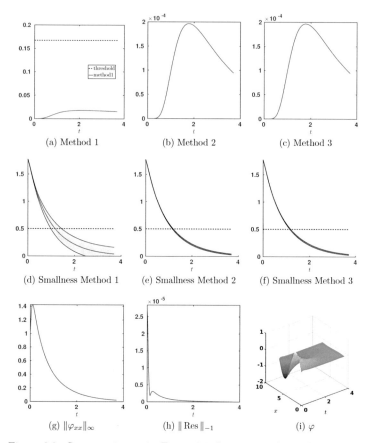

Figure 3.2: Same setting as in Figure 3.1, but now with smaller step-size $h = 10^{-6}$. Now Method 1 succeeds, too.

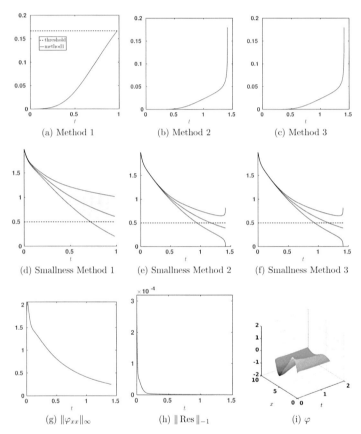

Figure 3.3: Initial value $u_0 = \sin(x) + 0.25\sin(2x)$, $N = 128$ Fourier modes and step-size $h = 10^{-6}$. All three methods fail both, the time and smallness criterion.

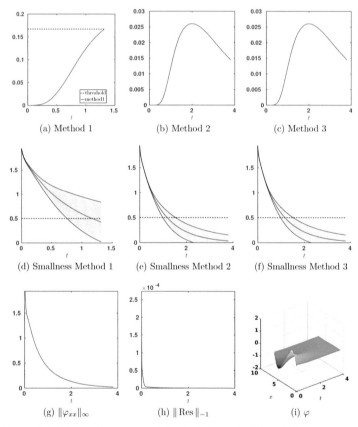

Figure 3.4: Initial value $u_0 = \sin(x) + 0.225\sin(2x)$, $N = 128$ Fourier modes and step-size $h = 10^{-6}$. Here only Method 1 fails.

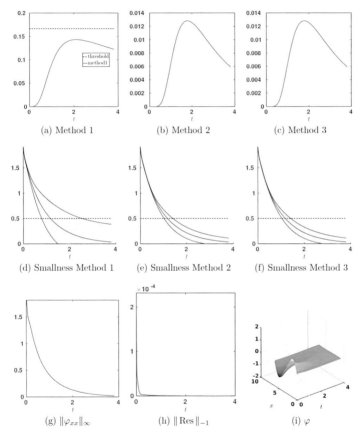

(a) Method 1   (b) Method 2   (c) Method 3

(d) Smallness Method 1   (e) Smallness Method 2   (f) Smallness Method 3

(g) $\|\varphi_{xx}\|_\infty$ (h) $\|\operatorname{Res}\|_{-1}$ (i) φ

Figure 3.5: Initial value $u_0 = \sin(x) + 0.2\sin(2x)$, $N = 128$ Fourier modes and step-size $h = 10^{-6}$. All three methods succeed the time and smallness criterion.

				Smallness			Time		
u_0	$T^* \approx$	N	h	M1	M2	M3	M1	M2	M3
$\sin(x)$	3.5	128	10^{-5}	–	1.48	1.48	1.57	✓	✓
$\sin(x)$	3.5	128	10^{-6}	1.46	1.19	1.19	✓	✓	✓
$\sin(x) + 0.25\sin(2x)$	3.7	128	10^{-6}	–	–	–	0.98	1.41	1.41
$\sin(x) + 0.225\sin(2x)$	3.7	128	10^{-6}	–	1.53	1.53	1.32	✓	✓
$\sin(x) + 0.2\sin(2x)$	3.6	128	10^{-6}	2.5	1.4	1.4	✓	✓	✓

Table 3.1: Summary of the simulations of this chapter. All values are rounded to fit into the table. A "–" in the "Smallness" columns means, that the smallness criterion was not met by the respective method. Else, we give the time when it was met. For the "Time" columns this turns around. If the time criterion was met, the respective method gets a "✓", else the time of the blowup / reaching the threshold is displayed.

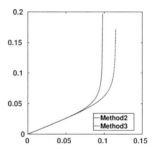

Figure 3.6: Artificial example for fixed $\| \operatorname{Res} \|_{-1}^2 = 0.5$ and $\|\varphi_{xx}\|_\infty^2 = 0.1$ to compare Method 2 with Method 3.

tive $\|\varphi_{xx}\|_\infty^2 = 0.1$, without using any numerical approximation (If these values are constant, we can calculate all integrals for the bounds exactly). The simulation shows, that Method 3 has indeed a better handling of large residuals. Usually, if we take a reasonable approximation φ (for interesting initial values), we would have a significantly smaller $\| \operatorname{Res} \|_{-1}^2$ compared to $\|\varphi_{xx}\|_\infty^2$ which is the reason why we could not see a difference in the previous simulations.

3.5   Conclusion

If we neglect that the simulations were not fully rigorous concerning the residual and interval arithmetic, all three methods have shown that they are capable of proving global existence of smooth solutions for the surface growth equation (1.1) for initial values which are not covered by a-priori results like the smallness criterion (Theorem 2.2).

Also, the examples of Figure 3.3, 3.4 and 3.5 show, that all three methods can react very sensitive to small alterations of the initial value. While this might be obvious for Method 1 with its threshold, the exponential nature of Methods 2 and 3 lead to a similar effect. In the end, Methods 2 and 3 look more promising than Method 1, because there were no examples for an initial value where the results by Method 1 were better than the other two methods (We have tested far more initial values than the few given here, but listing all of them would have no further benefit for comparing the methods).

Further, we were able to show that Method 3 is in fact superior to Method 2 due to its better handling of the residual. This is not really visible in our simulations because for a reasonable approximation φ, $\|\operatorname{Res}\|_{-1}$ is usually several orders of magnitude smaller than $\|\varphi_{xx}\|_{\infty}$. Therefore, we are confident that Method 3 and numerically improvements like more modes, smaller step-size or even a different numerical method are sufficient to control the residual as a reason for a blowup (Please note, that there are many other approaches conceivable than the three methods presented, for example restarting the first method or bounding the coefficients of the ODE and solving it directly are valid options, too, which are not part of this work). The remaining quantity that we have to control, and in fact the currently more influential one, is $\|\varphi_{xx}\|_{\infty}$ which we will take care of in the next chapter.

CHAPTER 4

Eigenvalue Estimate

We will now improve the bound (2.1), especially with regard to $\|\varphi_{xx}\|$ which enters the equation through the estimate of the (B) term. The aim is to find a better estimate for the combined (A) and (B) terms (which is basically the linearization of the nonlinear operator around the numerical approximation) by establishing an eigenvalue estimate that makes use of the finite dimensional estimate and a rigorous error bound. Note, that there are many theoretical results known (e.g. [Liu15]), but we have to find an estimate that is also easily and accurately computable.

4.1 Setting and Theorem

Using the setting from Section 1.2, we consider the linear operators

$$L_\varphi u = -\partial_x^4 u - 2\partial_x^2(\varphi_x u_x)$$

and

$$A_\varphi u = -\partial_x^4 u - 2\partial_x^3(\varphi_x u),$$

where $\varphi \in \mathcal{H}^4$ in space and H^1 in time is some arbitrary but fixed function like before.

We are interested in bounding the quadratic form

$$\lambda = \sup_{\|u_x\|=1} \langle \partial_x L_\varphi u, \partial_x u \rangle \tag{4.1}$$

31

in order to finally obtain a bound

$$\langle \partial_x L_\varphi u, \partial_x u \rangle \le \lambda \|u_x\|^2.$$

By substituting $v = u_x$ we immediately get

$$\lambda = \sup_{\|v\|=1} \langle A_\varphi v, v \rangle .$$

Define H_n as the $2n$-dimensional subspace spanned by e^{ix}, ..., e^{inx} and its complex conjugates e^{-ix}, ..., e^{-inx} (we can omit the constant mode due to our solution space $\mathcal{H}$). Denote by P_n the orthogonal projection onto H_n. Finally, we set

$$\lambda_n := \sup_{\|u\|=1} \langle P_n A_\varphi P_n u, u \rangle, \tag{4.2}$$

which is just the largest eigenvalue of the symmetric $2n \times 2n$ matrix $\frac{1}{2}(P_n A_\varphi + A^* P_n)$ (see Chapter B.2).

Obviously, as the supremum is over a larger set, it holds that

$$\lambda_n \le \lambda.$$

Now we want to bound λ from above.

For our first estimate (2.1) we used integration by parts, interpolation, and Poincare inequalities to obtain the following "worst case" estimate

$$\begin{aligned}
\langle A_\varphi u, u \rangle &= -\|u_{xx}\|^2 + 2 \int \varphi_x u u_{xxx} \; \mathrm{d}x \\
&= -\|u_{xx}\|^2 - 2 \int \varphi_{xx} u u_{xx} \; \mathrm{d}x - 2 \int \varphi_x u_x u_{xx} \; \mathrm{d}x \\
&= -\|u_{xx}\|^2 - 2 \int \varphi_{xx} u u_{xx} \; \mathrm{d}x + \int \varphi_{xx} u_x^2 \; \mathrm{d}x \\
&\le -\|u_{xx}\|^2 + 2\|\varphi_{xx}\|_\infty \|u\| \|u_{xx}\| + \|\varphi_{xx}\|_\infty \|u_x\|^2 \\
&\le -\|u_{xx}\|^2 + 3\|\varphi_{xx}\|_\infty \|u\| \|u_{xx}\| \\
&\le -\tfrac{1}{2}\|u_{xx}\|^2 + \tfrac{9}{2}\|\varphi_{xx}\|_\infty^2 \|u\|^2 \\
&\le -\tfrac{1}{2}\|u\|^2 + \tfrac{9}{2}\|\varphi_{xx}\|_\infty^2 \|u\|^2.
\end{aligned}$$

Thus

$$\lambda \le -\frac{1}{2} + \frac{9}{2}\|\varphi_{xx}\|_\infty^2 . \tag{4.3}$$

Instead, the following theorem shows an improved estimate by analyzing the quadratic form (4.1) separately for different mode ranges.

Theorem 4.1. *Let u be a solution to our surface growth equation (1.1), $\varphi \in H_n$ an arbitrary approximation and H_n, λ and λ_n be defined as in Section 4.1. Then, for*

$$n \geq \sqrt{2}C_\varphi = \sqrt{2}(2\|\varphi_{xxx}\|_\infty + 6\|\varphi_{xx}\|_\infty + 4\|\varphi_x\|_\infty)$$

it holds that

$$\lambda_n \leq \lambda \leq \lambda_n + \frac{1}{2}\max\left\{2C_\varphi^2 \frac{9\|\varphi_{xx}\|_\infty^2 - 2\lambda_n}{n^2} \ , \ 9\|\varphi_{xx}\|_\infty^2 + 2\lambda_n - \frac{1}{2}n^4\right\}.$$

4.2 Proof of the Theorem

As a preparation, we split $u = p + q$, where $p \in H_n$ and $q \perp H_n$. Thus

$$\lambda = \sup_{\|u\|=1} \langle A_\varphi u, u \rangle$$
$$= \sup_{\|p\|^2+\|q\|^2=1} \left\{ \langle A_\varphi p, p \rangle + \langle A_\varphi p, q \rangle + \langle A_\varphi q, p \rangle + \langle A_\varphi q, q \rangle \right\}.$$

Now, we will treat these scalar products separately, where we will denote with "low modes" the parts only depending on p and with "high modes" everything solely depending on q.

Low modes

First, notice that analogous to the brute force estimate

$$\langle A_\varphi p, p \rangle \leq -\frac{1}{2}\|p_{xx}\|^2 + \frac{9}{2}\|\varphi_{xx}\|_\infty^2 \|p\|^2.$$

Second, it holds by definition of λ_n, as $p \in H_n$

$$\langle A_\varphi p, p \rangle \leq \lambda_n \|p\|^2.$$

Thus in summary, we get

$$\langle A_\varphi p, p \rangle \leq (1 - \eta_n)\lambda_n \|p\|^2 - \frac{1}{2}\eta_n \|p_{xx}\|^2 + \frac{9}{2}\eta_n \|\varphi_{xx}\|_\infty^2 \|p\|^2$$

for some $\eta_n \in [0, 1]$, that we will fix later.

Mixed terms

For the mixed terms we use

$$\|p\| \leq \|p_x\| \leq \|p_{xx}\| \quad \text{and} \quad \|q\| \leq \frac{1}{n}\|q_x\| \leq \frac{1}{n^2}\|q_{xx}\|$$

to obtain (any derivatives of p and q are still orthogonal in $\mathcal{H}$)

$$
\begin{aligned}
\langle A_\varphi p, q \rangle &= -2 \int (\varphi_x p)_{xxx} q \; \mathrm{d}x \\
&= 2 \int (\varphi_x p)_{xx} q_x \; \mathrm{d}x \\
&= 2 \int (\varphi_{xxx} p + 2\varphi_{xx} p_x + \varphi_x p_{xx}) q_x \; \mathrm{d}x \\
&\leq 2\|q_x\| \cdot \left(\|\varphi_{xxx}\|_\infty \|p\| + 2\|\varphi_{xx}\|_\infty \|p_x\| + \|\varphi_x\|_\infty \|p_{xx}\| \right) \\
&\leq C_\varphi^{(1)} \frac{1}{n} \|q_{xx}\| \|p_{xx}\|
\end{aligned}
$$

with $C_\varphi^{(1)} = 2\|\varphi_{xxx}\|_\infty + 4\|\varphi_{xx}\|_\infty + 2\|\varphi_x\|_\infty]$ and

$$
\begin{aligned}
\langle A_\varphi q, p \rangle &= 2 \int (\varphi_x q) p_{xxx} \; \mathrm{d}x \\
&= -2 \int (\varphi_x q)_x p_{xx} \; \mathrm{d}x \\
&\leq 2\|p_{xx}\| \cdot \left(\|\varphi_{xx}\|_\infty \|q\| + \|\varphi_x\|_\infty \|q_x\| \right) \\
&\leq C_\varphi^{(2)} \frac{1}{n} \|q_{xx}\| \|p_{xx}\|
\end{aligned}
$$

with $C_\varphi^{(2)} = 2\|\varphi_{xx}\|_\infty + 2\|\varphi_x\|_\infty$. Further, we define

$$C_\varphi = C_\varphi^{(1)} + C_\varphi^{(2)} = 2\|\varphi_{xxx}\|_\infty + 6\|\varphi_{xx}\|_\infty + 4\|\varphi_x\|_\infty \; .$$

High modes

Finally, for the high modes we use the rough "worst case" estimate (4.3),

$$\langle A_\varphi q, q \rangle \leq -\frac{1}{2}\|q_{xx}\|^2 + \frac{9}{2}\|\varphi_{xx}\|_\infty^2 \|q\|^2,$$

but we will later apply the improved Poincare inequality

$$\|q\| \leq \frac{1}{n}\|q_x\| \quad \forall q \perp H_n. \tag{4.4}$$

Summary

Combining all estimates, we obtain (using $ab \leq \frac{1}{2}a^2 + \frac{1}{2}b^2$ and eliminating p_{xx} terms)

$$\langle A_\varphi u, u \rangle = \langle A_\varphi p, p \rangle + \langle A_\varphi p, q \rangle + \langle A_\varphi q, p \rangle + \langle A_\varphi q, q \rangle$$

$$\leq (1 - \eta_n)\lambda_n \|p\|^2 - \frac{1}{2}\eta_n \|p_{xx}\|^2 + \frac{9}{2}\eta_n \|\varphi_{xx}\|_\infty^2 \|p\|^2$$

$$+ C_\varphi \frac{1}{n} \|q_{xx}\| \|p_{xx}\|$$

$$- \frac{1}{2}\|q_{xx}\|^2 + \frac{9}{2}\|\varphi_{xx}\|_\infty^2 \|q\|^2$$

$$\leq (1 - \eta_n)\lambda_n \|p\|^2 + \frac{9}{2}\eta_n \|\varphi_{xx}\|_\infty^2 \|p\|^2$$

$$+ \frac{1}{2}\left(\frac{C_\varphi^2}{n^2\eta_n} - 1\right)\|q_{xx}\|^2 + \frac{9}{2}\|\varphi_{xx}\|_\infty^2 \|q\|^2$$

In order to apply the improved Poincare inequality (4.4) for q, we define

$$\eta_n := 2\frac{C_\varphi^2}{n^2} \text{ and thus } n \geq \sqrt{2}C_\varphi \text{ to assert } \eta_n \leq 1$$

and obtain

$$\langle A_\varphi u, u \rangle \leq \left[(1 - \eta_n)\lambda_n + \frac{9}{2}\eta_n \|\varphi_{xx}\|_\infty^2\right]\|p\|^2$$

$$+ \frac{1}{2}\left[9\|\varphi_{xx}\|_\infty^2 - \frac{1}{2}n^4\right]\|q\|^2$$

which proves our main theorem

$$\lambda = \sup_{\|u\|=1} \langle A_\varphi u, u \rangle = \sup_{\|p\|^2 + \|q\|^2 = 1} \langle A_\varphi u, u \rangle$$

$$\leq \max\left\{\left[(1 - \eta_n)\lambda_n + \frac{9}{2}\eta_n \|\varphi_{xx}\|_\infty^2\right], \frac{1}{2}\left[9\|\varphi_{xx}\|_\infty^2 - \frac{1}{2}n^4\right]\right\}$$

$$= \lambda_n + \frac{1}{2}\max\left\{\eta_n[9\|\varphi_{xx}\|_\infty^2 - 2\lambda_n], 9\|\varphi_{xx}\|_\infty^2 + 2\lambda_n - \frac{1}{2}n^4\right\}.$$

$\square$

4.3 Comparison with the Previous Estimate

Now, we want to give a rough comparison between the new and the old estimate of the quadratic form. Recall the previous worst case estimate (4.3),

$$\lambda \leq -\frac{3}{4} + 9\|\varphi_{xx}\|_\infty^2,$$

that we used for Methods 1 - 3 (and will use for Method 4 in the following chapter). We can rewrite Theorem 4.1 in the following way:

$$\lambda \leq \lambda_n + \frac{1}{2}\max\left\{2C_\varphi^2\frac{9\|\varphi_{xx}\|_\infty^2 - 2\lambda_n}{n^2} \ , \ 9\|\varphi_{xx}\|_\infty^2 + 2\lambda_n - \frac{1}{2}n^4\right\}$$

$$= \max\left\{\frac{9C_\varphi^2\|\varphi_{xx}\|_\infty^2}{n^2} + \left(1 - \frac{2C_\varphi^2}{n^2}\right)\lambda_n \ , \ \frac{9}{2}\|\varphi_{xx}\|_\infty^2 + 2\lambda_n - \frac{1}{4}n^4\right\}$$

$$= \frac{9}{2}\|\varphi_{xx}\|_\infty^2 + \max\left\{\left(\frac{9C_\varphi^2}{n^2} - \frac{9}{2}\right)\|\varphi_{xx}\|_\infty^2 + \left(1 - \frac{2C_\varphi^2}{n^2}\right)\lambda_n \ , \ 2\lambda_n - \frac{1}{4}n^4\right\}$$

and as $\frac{C_\varphi^2}{n^2} \leq \frac{1}{2}$

$$\leq \frac{9}{2}\|\varphi_{xx}\|_\infty^2 + \max\left\{0 \ , \ 2\lambda_n - \frac{1}{4}n^4\right\}.$$

Thus, as long as the worst case estimate is finite, our new estimate is at least less than roughly a half of it, as the max tends to zero with increasing n (as λ_n is bounded by λ which is in itself bounded by the worst case estimate that is independent of n) and $-\frac{3}{4}$ is negligible for all interesting cases. In fact, most of the time we have far better results than just a factor of $\frac{1}{2}$ because $\frac{C_\varphi^2}{n^2} \ll \frac{1}{2}$.

4.4   Application to the Surface Growth Equation

Let us denote the eigenvalue bound from Theorem 4.1 with $\tilde{\lambda}$. If we want to incorporate this result into our framework, we have to consider, that in order to control the (C) and (D) terms, we need some part of the (A) term of

$$\frac{1}{2}\partial_t\|d_x\|^2 = \underbrace{\langle d_{xx}, d_{xxxx} + 2(d_x\varphi_x)_{xx}\rangle}_{A+B} + \underbrace{\langle d_{xx}, (d_x{}^2)_{xx} + \text{Res}\rangle}_{C+D}.$$

Therefore we split the first term into two parts

$$\frac{1}{2}\partial_t\|d_x\|^2 = (1-\delta)\langle d_{xx}, d_{xxxx} + 2(d_x\varphi_x)_{xx}\rangle + \delta\langle d_{xx}, d_{xxxx} + 2(d_x\varphi_x)_{xx}\rangle$$
$$+ \langle d_{xx}, (d_x{}^2)_{xx} + \text{Res}\rangle.$$

Now we can bound the first term with our new method and the remaining parts like before in (2.1). If we do not fix the constants used in the Young inequalities, we have

$$A = -\|d_{xxx}\|^2$$

$$|B| \leq \varepsilon_B \|d_{xxx}\|^2 + \frac{9}{4\varepsilon_B} \|d_x\|^2 \|\varphi_{xx}\|_\infty^2$$

$$|C| \leq \varepsilon_C \|d_{xxx}\|^2 + \frac{(\frac{4}{7}\varepsilon_C)^{-7}}{4} \|d_x\|^{10}$$

$$|D| \leq \varepsilon_D \|d_{xxx}\|^2 + \frac{1}{4\varepsilon_D} \|\operatorname{Res}\|_{-1}^2,$$

where we can set all $\varepsilon_{\{B,C,D\}} > 0$ arbitrary small.

In this case, our differential inequality is

$$\frac{1}{2}\partial_t\|d_x\|^2 \leq (1-\delta)\tilde{\lambda}\|d_x\|^2 + \frac{9}{4\varepsilon_B}\delta\|d_x\|^2\|\varphi_{xx}\|_\infty^2 + \frac{(\frac{4}{7}\varepsilon_C)^{-7}}{4}\|d_x\|^{10}$$

$$+ \frac{1}{4\varepsilon_D}\|\operatorname{Res}\|_{-1}^2 + \left(\delta\varepsilon_B + \varepsilon_C + \varepsilon_D - \delta\right)\|d_{xxx}\|^2,$$

where $\varepsilon_{\{B,C,D\}} > 0$ and $\delta \in (0,1)$. By substituting $\varepsilon_{\{C,D\}} := \delta\varepsilon_{\{C,D\}}$, this is equivalent to

$$\frac{1}{2}\partial_t\|d_x\|^2 \leq (1-\delta)\tilde{\lambda}\|d_x\|^2 + \frac{9}{4\varepsilon_B}\delta\|d_x\|^2\|\varphi_{xx}\|_\infty^2 + \frac{(\frac{4}{7}\delta\varepsilon_C)^{-7}}{4}\|d_x\|^{10}$$

$$+ \frac{1}{4\delta\varepsilon_D}\|\operatorname{Res}\|_{-1}^2 + \delta\left(\varepsilon_B + \varepsilon_C + \varepsilon_D - 1\right)\|d_{xxx}\|^2,$$

where $\varepsilon_{\{B,C,D\}} > 0$ and $\delta \in (0,1)$. Next, we set $\varepsilon_B + \varepsilon_C + \varepsilon_D = 1$ to remove the last term, and therefore, we are now interested in the minimum of

$$\frac{1}{2}\partial_t\|d_x\|^2 \leq (1-\delta)\tilde{\lambda}\|d_x\|^2 + \frac{9\delta}{4\varepsilon_B}\|d_x\|^2\|\varphi_{xx}\|_\infty^2 + \frac{7^7}{4^8(\delta\varepsilon_C)^7}\|d_x\|^{10}$$

$$+ \frac{1}{4\delta\varepsilon_D}\|\operatorname{Res}\|_{-1}^2 \tag{4.5}$$

under the constraints $\varepsilon_{\{B,C,D\}} > 0$, $\sum_{k\in\{B,C,D\}} \varepsilon_k = 1$, $\delta \in (0,1)$.

Unfortunately, it is not so easy to determine a rigorous global minimum for this problem. We first rewrite the function and condense all constants

$$f(x) = (1-x_1)c_1 + \frac{x_1}{x_2}c_2 + (x_1x_3)^{-7}c_3 + (x_1x_4)^{-1}c_4$$

with $c_{2,3,4} \geq 0$ and the constraints of

$$0 < x_1 < 1$$
$$x_{2,3,4} > 0$$
$$x_2 + x_3 + x_4 = 1.$$

Now, we look at the function independent of x_1, by moving x_1 to the constants and proceed to solve the minimization only for x_2, x_3, x_4

$$f(x) = \tilde{c}_1 + x_2^{-1}\tilde{c}_2 + x_3^{-7}\tilde{c}_3 + x_4^{-1}\tilde{c}_4$$
$$1 = x_2 + x_3 + x_4$$
$$0 < x_2, x_3, x_4.$$

Lagrange multipliers now yield

$$\begin{pmatrix} -x_2^{-2}\tilde{c}_2 \\ -x_3^{-8}7\tilde{c}_3 \\ -x_4^{-2}\tilde{c}_4 \\ x_2 + x_3 + x_4 \end{pmatrix} \stackrel{!}{=} \begin{pmatrix} \lambda \\ \lambda \\ \lambda \\ 1 \end{pmatrix},$$

and thus as $x_k > 0$ and $\tilde{c}_{2,3,4} > 0$, and therefore $\lambda < 0$, we have

$$x_2 = \sqrt{-\frac{\tilde{c}_2}{\lambda}}, \qquad x_3 = \sqrt[8]{-\frac{7\tilde{c}_3}{\lambda}}, \qquad x_4 = \sqrt{-\frac{\tilde{c}_4}{\lambda}},$$
$$\Rightarrow 1 = \sqrt{-\frac{\tilde{c}_2}{\lambda}} + \sqrt{-\frac{\tilde{c}_4}{\lambda}} + \sqrt[8]{-\frac{7\tilde{c}_3}{\lambda}}$$
$$= \frac{1}{\sqrt{|\lambda|}}(\sqrt{\tilde{c}_2} + \sqrt{\tilde{c}_4}) + \frac{1}{\sqrt[8]{|\lambda|}}\sqrt[8]{7\tilde{c}_3}.$$

This is an equation of type

$$ct^4 + t - 1 = 0,$$

which we could solve by Ferrari's method for quartic equations (see e.g. [Neu65] or any other basic algebra book). Given the length of Ferrari's formula and the fact that we further had to take into account that now all constants depend on x_1 (which leads to another increase in complexity because of the derivative), we will not pursuit this ansatz as its cost-benefit ratio is not good enough. Luckily, we can not do anything wrong here that breaks the rigorosity of our calculations, as valid parameter combinations just might not be optimal. Therefore, we will just use MATLAB's nonlinear optimization solver to find a approximate local minimum and update it after a given time interval (we could do this in every step, but given that the step-size is small enough and the data is continuous, this is not necessary and would just cost us lots of computational time).

CHAPTER 5

Numerical Preparations

From our numerical algorithm we obtain discrete values of φ for times $0 = t_0, t_1, \ldots, t_n$. In order to calculate quantities like $A(t)$, $\tilde{b}(t)$, and $\tilde{f}(t)$, which occur in our bounds that we derived in Chapters 2 and 4, we have to define φ in between these grid points. Again we have to make a trade off between accuracy and feasibility, as there are multiple interpolation methods applicable.

In our case, we will set φ as the piecewise linear interpolation of our discrete values

$$\varphi(x, t_0 + t) = (1 - \tfrac{t}{h})\varphi(x, t_0) + \tfrac{t}{h}\varphi(x, t_1)$$
$$= \tfrac{t}{h}(\varphi(x, t_1) - \varphi(x, t_0)) + \varphi(x, t_0) \qquad t \in [0, h], \ h = t_1 - t_0.$$

Following this ansatz, we have to derive computable quantities like $\| \operatorname{Res} \|_{-1}$ and $A(t)$ for every time $t \in [t_0, t_1]$, or at least obtain upper bounds, using only known values at time t_0 or t_1.

5.1   Norm of the Residual

For the residual, we get by using the definition and applying the linear interpolation

$$\operatorname{Res} \varphi(x, t_0 + t) = \partial_t \varphi(x, t_0 + t) + \partial_x^4 \varphi(x, t_0 + t) + \partial_x^2 (\partial_x \varphi(x, t_0 + t))^2$$
$$= \tfrac{1}{h}(\varphi(x, t_1) - \varphi(x, t_0))$$

$$+ \tfrac{t}{h}\partial_x^4(\varphi(x,t_1) - \varphi(x,t_0)) + \partial_x^4\varphi(x,t_0)$$
$$+ \partial_x^2(\partial_x\varphi(x,t_0))^2$$
$$+ 2\tfrac{t}{h}\partial_x^2(\partial_x(\varphi(x,t_1) - \varphi(x,t_0)) \cdot \partial_x\varphi(x,t_0))$$
$$+ \tfrac{t^2}{h^2}\partial_x^2(\partial_x(\varphi(x,t_1) - \varphi(x,t_0)))^2$$
$$= \tfrac{1}{h}(\varphi(x,t_1) - \varphi(x,t_0)) + \partial_x^4\varphi(x,t_0) + \partial_x^2(\partial_x\varphi(x,t_0))^2$$
$$+ \tfrac{t}{h}[\partial_x^4(\varphi(x,t_1) - \varphi(x,t_0))$$
$$\qquad + 2\partial_x^2(\partial_x(\varphi(x,t_1) - \varphi(x,t_0)) \cdot \partial_x\varphi(x,t_0))]$$
$$+ \tfrac{t^2}{h^2}\partial_x^2(\partial_x(\varphi(x,t_1) - \varphi(x,t_0)))^2.$$

And therefore, we have for the spatial primitive

$$\int \mathrm{Res}\,\varphi(x,t_0 + t) \,\mathrm{d}x = \tfrac{1}{h}(\Phi(x,t_1) - \Phi(x,t_0)) + \partial_x^3\varphi(x,t_0) + \partial_x(\partial_x\varphi(x,t_0))^2$$
$$+ t\tfrac{1}{h}[\partial_x^3(\varphi(x,t_1) - \varphi(x,t_0))$$
$$\qquad + 2\partial_x(\partial_x(\varphi(x,t_1) - \varphi(x,t_0)) \cdot \partial_x\varphi(x,t_0))]$$
$$+ t^2\tfrac{1}{h^2}\partial_x(\partial_x(\varphi(x,t_1) - \varphi(x,t_0)))^2$$
$$=: a(x) + tb(x) + t^2c(x),$$

where we use the following abbreviations

$$a(x) = \tfrac{1}{h}(\Phi(x,t_1) - \Phi(x,t_0)) + \partial_x^3\varphi(x,t_0) + \partial_x(\partial_x\varphi(x,t_0))^2,$$
$$b(x) = \tfrac{1}{h}[\partial_x^3(\varphi(x,t_1) - \varphi(x,t_0)) + 2\partial_x(\partial_x(\varphi(x,t_1) - \varphi(x,t_0)) \cdot \partial_x\varphi(x,t_0))],$$
$$c(x) = \tfrac{1}{h^2}\partial_x(\partial_x(\varphi(x,t_1) - \varphi(x,t_0)))^2$$

and

$$\Phi(x) = \int \varphi(x) \,\mathrm{d}x.$$

The square is now given by a simple multiplication

$$\left(\int \mathrm{Res}\,\varphi(x,t_0 + t) \,\mathrm{d}x\right)^2 = a(x)^2$$
$$+ t2a(x)b(x)$$
$$+ t^2(b(x)^2 + 2a(x)c(x))$$
$$+ t^3 2b(x)c(x)$$
$$+ t^4 c(x)^2.$$

Finally, we have

$$
\begin{aligned}
\| \operatorname{Res} \varphi(x, t_0 + t)\|_{-1}^2 = &\int_0^{2\pi} a(x)^2 \, \mathrm{d}x \\
&+ 2t \int_0^{2\pi} a(x)b(x) \, \mathrm{d}x \\
&+ t^2 \int_0^{2\pi} b(x)^2 + 2a(x)c(x) \, \mathrm{d}x \\
&+ 2t^3 \int_0^{2\pi} b(x)c(x) \, \mathrm{d}x \\
&+ t^4 \int_0^{2\pi} c(x)^2 \, \mathrm{d}x
\end{aligned}
\tag{5.1}
$$

or in a shorter form that we will use later

$$
= C_0 + tC_1 + t^2 C_2 + t^3 C_3 + t^4 C_4.
$$

Accordingly, we have

$$
\begin{aligned}
\int_0^h \| \operatorname{Res} \varphi(x, t_0 + t)\|_{-1}^2 \, \mathrm{d}t = &h \int_0^{2\pi} a(x)^2 \, \mathrm{d}x \\
&+ h^2 \int_0^{2\pi} a(x)b(x) \, \mathrm{d}x \\
&+ \tfrac{1}{3} h^3 \int_0^{2\pi} b(x)^2 + 2a(x)c(x) \, \mathrm{d}x \\
&+ \tfrac{1}{2} h^4 \int_0^{2\pi} b(x)c(x) \, \mathrm{d}x \\
&+ \tfrac{1}{5} h^5 \int_0^{2\pi} c(x)^2 \, \mathrm{d}x,
\end{aligned}
\tag{5.2}
$$

what we will also use later.

Note, that we can calculate the integrals of the right-hand side for both equations (5.1) and (5.2) using the L^2-scalar product accurately and efficiently as our simulation uses Fourier transformations.

5.2 Rigorous Computable Bound for Method 3

Now we want to derive a computable upper bound for Method 3 as stated in Theorem 3.3. First, we define

$$
T := e^{A(h)} \left(z(0) + \int_0^h \tilde{f}(s) \, \mathrm{d}s \right)
$$

$$= e^{A(h)}z(0) + 2\int_0^h e^{A(h)-A(s)} \|\operatorname{Res}\|_{-1}^2 \ ds.$$

Therefore, our bound for Method 3 on one step of size h is given by

$$T \times \left(1 - 4T^4 \times \frac{7^7}{2}\int_0^h e^{4(A(s)-A(h))} \ ds\right)^{-1/4}. \tag{5.3}$$

To get an upper bound for (5.3), we need an upper bound on T itself, and an upper bound for the exponential term $\int_0^h e^{4(A(s)-A(h))} \ ds$. Therefore, we need an upper bound for $A(t)$, where we have

$$
\begin{aligned}
A(t) &= -\frac{1}{2}t + 18\int_0^t \|\varphi_{xx}(t_0+s)\|_{L^\infty}^2 \ ds \\
&\leq -\frac{1}{2}t + 18\int_0^t ((1-\tfrac{s}{h})\|\varphi_{xx}(t_0)\|_{L^\infty} + \tfrac{s}{h}\|\varphi_{xx}(t_1)\|_{L^\infty})^2 \ ds \\
&= -\frac{1}{2}t + 18\Big((t - \frac{t^2}{h} + \frac{t^3}{3h^2})\|\varphi_{xx}(t_0)\|_\infty^2 \\
&\quad + 2\Big(\frac{t^2}{2h} - \frac{t^3}{3h^2}\Big)\|\varphi_{xx}(t_0)\|_\infty \|\varphi_{xx}(t_1)\|_\infty \\
&\quad + \frac{t^3}{3h^2}\|\varphi_{xx}(t_1)\|_\infty^2\Big),
\end{aligned}
\tag{5.4}
$$

that we can use for

$$
\begin{aligned}
A(h) - A(s) &= \frac{1}{2}(s-h) + 18\int_s^h \|\varphi_{xx}\|_\infty^2 \ d\tau \\
&\leq \frac{1}{2}(s-h) \\
&\quad + 18\Big(\Big(h - s - \frac{h^2+s^2}{h} + \frac{h^3-s^3}{3h^2}\Big)\|\varphi_{xx}(t_0)\|_\infty^2 \\
&\quad + 2\Big(\frac{h^2-s^2}{2h} - \frac{h^3+s^3}{3h^2}\Big)\|\varphi_{xx}(t_0)\|_\infty \|\varphi_{xx}(t_1)\|_\infty \\
&\quad + \frac{h^3-s^3}{3h^2}\|\varphi_{xx}(t_1)\|_\infty^2\Big).
\end{aligned}
\tag{5.5}
$$

Further, the upper bound on T consists of two parts, an upper bound for $e^{A(h)}z(0)$, what we basically handled before, and an upper bound for $\int_0^h e^{A(h)-A(s)}\|\operatorname{Res}\|_{-1}^2 \ ds$. The first part can be treated by using (5.4),

$$
\begin{aligned}
e^{A(h)}z(0) &\leq \exp\Big\{-\frac{1}{2}h + 18\Big(\frac{h}{3}\|\varphi_{xx}(t_0)\|_\infty^2 \\
&\quad + \frac{h}{3}\|\varphi_{xx}(t_0)\|_\infty\|\varphi_{xx}(t_1)\|_\infty + \frac{h}{3}\|\varphi_{xx}(t_1)\|_\infty^2\Big)\Big\}z(0)
\end{aligned}
\tag{5.6}
$$

and for the second part we get by (5.1) and (5.5)

$$
\int_0^h e^{A(h)-A(s)} \|\operatorname{Res}\|_{-1}^2 \ \mathrm{d}s
$$

$$
\leq \int_0^h \exp\left\{ \frac{1}{2}(s-h) \right.
$$

$$
+ 18\left(\left(h - s - \frac{h^2+s^2}{h} + \frac{h^3-s^3}{3h^2} \right) \|\varphi_{xx}(t_0)\|_\infty^2 \right.
$$

$$
+ 2\left(\frac{h^2-s^2}{2h} - \frac{h^3+s^3}{3h^2} \right) \|\varphi_{xx}(t_0)\|_\infty \|\varphi_{xx}(t_1)\|_\infty
$$

$$
\left. \left. + \frac{h^3-s^3}{3h^2} \|\varphi_{xx}(t_1)\|_\infty^2 \right) \right\}
$$

$$
\times \left\{ C_0 + sC_1 + s^2C_2 + s^3C_3 + s^4C_4 \right\} \ \mathrm{d}s.
$$

At this point we have several options how to estimate this expression. A not so sophisticated option is the following, where we basically integrate from 0 instead of s in the $A(h) - A(s)$ term

$$
\int_0^h e^{A(h)-A(s)} \|\operatorname{Res}\|_{-1}^2 \ \mathrm{d}s
$$

$$
\leq \exp\left\{ 18\left(\frac{h}{3}\|\varphi_{xx}(t_0)\|_\infty^2 + \frac{h}{3}\|\varphi_{xx}(t_0)\|_\infty \|\varphi_{xx}(t_1)\|_\infty \right.\right.
$$

$$
\left.\left. + \frac{h}{3}\|\varphi_{xx}(t_1)\|_\infty^2 \right) \right\} \tag{5.7}
$$

$$
\times \int_0^h C_0 + sC_1 + s^2C_2 + s^3C_3 + s^4C_4 \ \mathrm{d}s.
$$

The remaining integral is given by (5.2).

Additionally, as mentioned before, we need an upper bound for $\int_0^h e^{A(s)-A(h)} \ \mathrm{d}s$ (mind the different order of $A(s)$ and $A(h)$):
Therefore, we first take care of the exponent by a simple integral estimate using the minimum of the integrand. With

$$
A(s) - A(h) = \frac{1}{2}(h-s) - 18\int_s^h \|\varphi_{xx}\|_\infty^2 \ \mathrm{d}\tau
$$

$$
\leq \frac{1}{2}(h-s) - 18(h-s) \min_{t\in[t_0,t_1]} \|\varphi_{xx}\|_\infty^2
$$

$$
= (h-s)\left(\frac{1}{2} - 18 \min_{t\in[t_0,t_1]} \|\varphi_{xx}\|_\infty^2 \right)
$$

we have

$$\int_0^h e^{4(A(s)-A(h))} \, ds \leq \int_0^h \exp\left((h-s)\underbrace{\left(2 - 72\min_{t\in[t_0,t_1]}\|\varphi_{xx}\|_\infty^2\right)}_{C}\right) ds$$

$$= e^{Ch}\int_0^h e^{-Cs} \, ds = \frac{e^{Ch}-1}{C} \tag{5.8}$$

$$= \frac{e^{h(2-72\min_{t\in[t_0,t_1]}\|\varphi_{xx}\|_\infty^2)}-1}{2-72\min_{t\in[t_0,t_1]}\|\varphi_{xx}\|_\infty^2}$$

To sum up, our rigorously computable upper bound for Method 3, which we will define as Method 4 to separate it from the heuristic simulations of Chapter 3, is given by

Theorem 5.1 (Method 4). *Given the same setting as in Theorem 3.3. Let $t_{i+1}-t_i = h$, then it holds that*

$$z(t_{i+1}) \leq T \times \left(1 - 4T^4 \times \frac{7^7}{2}D\right)^{-1/4},$$

where

$$T = \exp\left\{-\frac{1}{2}h + 18\left(\frac{h}{3}\|\varphi_{xx}(t_i)\|_\infty^2 + \frac{h}{3}\|\varphi_{xx}(t_i)\|_\infty\|\varphi_{xx}(t_{i+1})\|_\infty\right.\right.$$
$$\left.\left.+ \frac{h}{3}\|\varphi_{xx}(t_{i+1})\|_\infty^2\right)\right\}z(t_i)$$
$$+ 2\exp\left\{18\left(\frac{h}{3}\|\varphi_{xx}(t_i)\|_\infty^2 + \frac{h}{3}\|\varphi_{xx}(t_i)\|_\infty\|\varphi_{xx}(t_{i+1})\|_\infty\right.\right.$$
$$\left.\left.+ \frac{h}{3}\|\varphi_{xx}(t_{i+1})\|_\infty^2\right)\right\}$$
$$\times \left(h\int_0^{2\pi} a(x)^2 \, dx + h^2 \int_0^{2\pi} a(x)b(x) \, dx\right.$$
$$+ \tfrac{1}{3}h^3\int_0^{2\pi} b(x)^2 + 2a(x)c(x) \, dx + \tfrac{1}{2}h^4\int_0^{2\pi} b(x)c(x) \, dx$$
$$\left.+ \tfrac{1}{5}h^5\int_0^{2\pi} c(x)^2 \, dx\right),$$
$$D = \frac{e^{h(2-72\min_{t\in[t_0,t_1]}\|\varphi_{xx}\|_\infty^2)}-1}{2-72\min_{t\in[t_0,t_1]}\|\varphi_{xx}\|_\infty^2}$$

and

$$a(x) = \tfrac{1}{h}(\Phi(x,t_{i+1}) - \Phi(x,t_i)) + \partial_x^3\varphi(x,t_i) + \partial_x(\partial_x\varphi(x,t_i))^2,$$

$b(x) = \frac{1}{h}[\partial_x^3(\varphi(x,t_{i+1}) - \varphi(x,t_i)) + 2\partial_x(\partial_x(\varphi(x,t_{i+1}) - \varphi(x,t_i))\partial_x\varphi(x,t_i))],$

$c(x) = \frac{1}{h^2}\partial_x(\partial_x(\varphi(x,t_{i+1}) - \varphi(x,t_i)))^2,$

$\Phi(x) = \int \varphi(x)\, dx.$

Proof. Apply the previous estimates (5.2), (5.6), (5.7), (5.8) and Theorem 3.3 to the differential inequality (5.3). $\qquad\qquad\qquad\qquad\qquad\square$

5.3 Rigorous Computable Bound including the Eigenvalue Estimate

How it does not work

(To reveal the outcome right away, this ansatz does not work here, as we will not be able to make one crucial estimate. We will do it nevertheless, to show what can be done and what not and why we have to do the estimates that we will do later.)

We will now try to follow the same procedure as in the previous section, but including the eigenvalue estimate from Theorem 4.1. The result for the Residual (5.1) is still valid, but we have to start with a new ODE:

$$\partial_t\|d_x\|^2 \leq \frac{7^7 \cdot 2}{4^8(\delta\varepsilon_C)^7}\|d_x\|^{10} + 2\Big((1-\delta)\lambda + \frac{9\delta}{4\varepsilon_B}\|\varphi_{xx}\|_\infty^2\Big)\|d_x\|^2 \tag{5.9}$$
$$+ \frac{1}{2\delta\varepsilon_D}\|\operatorname{Res}\|_{-1}^2$$

If we now apply Method 3 (Theorem 3.3), we have

$$\|d_x\|^2 \leq T \times \Big(1 - 4 \times T^4\frac{7^7\cdot 2}{4^8(\delta\varepsilon_C)^7}\int_0^h \exp\big(4(A(s) - A(h))\big)\ ds\Big)^{-1/4},$$

where

$$T := e^{A(h)}z(0) + \frac{1}{2\delta\varepsilon_D}\int_0^h e^{A(h)-A(s)}\|\operatorname{Res}(s)\|_{-1}^2\ ds$$

and

$$A(t) = 2\int_0^t (1-\delta)\lambda(s) + \frac{9\delta}{4\varepsilon_B}\|\varphi_{xx}(s)\|_\infty^2\ ds.$$

In addition to the estimates from the previous section, we have to take into account that our eigenvalue estimate λ is also time dependent and we have to apply the linear interpolation of φ to it. This also leads to more or less small alterations of the previous estimates.

Although our operator $A_\varphi u = -\partial_x^4 u - 2\partial_x^3(\varphi_x u)$ is not linear in φ, we have for $t_0 - t_1 = h$ and $t \in (0, h)$

$$A_{\varphi(t_0+t)}u = -\partial_x^4 u - 2\partial_x^3\Big(\big((1 - \tfrac{t}{h})\varphi_x(t_0) + \tfrac{t}{h}\varphi_x(t_1)\big)u\Big)$$

$$= (1 - \tfrac{t}{h})A_{\varphi(t_0)}u + \tfrac{t}{h}A_{\varphi(t_1)}u$$

and therefore

$$\lambda(t_0 + t) = \sup_{\|u\|=1} \langle A_{\varphi(t_0+t)}u, u\rangle = \sup_{\|u\|=1}\Big(\langle(1 - \tfrac{t}{h})A_{\varphi(t_0)}u, u\rangle + \langle\tfrac{t}{h}A_{\varphi(t_1)}u, u\rangle\Big)$$

$$\leq (1 - \tfrac{t}{h})\sup_{\|u\|=1}\langle A_{\varphi(t_0)}u, u\rangle + \tfrac{t}{h}\sup_{\|u\|=1}\langle A_{\varphi(t_1)}u, u\rangle$$

$$= (1 - \tfrac{t}{h})\lambda(t_0) + \tfrac{t}{h}\lambda(t_1) =: \lambda^*(t_0 + t).$$

This allows us to bound our eigenvalue estimate between two grid points.

We now proceed as in the previous section by setting up an upper bound for $A(t)$, which will then lead to an upper bound on T. The first estimates are nothing special and can be done as before. The new term involving the eigenvalue estimate is not hard to handle here.
For $A(t)$, we have

$$A(t) = 2\int_0^t (1-\delta)\lambda(t_0 + s) + \frac{9\delta}{4\varepsilon_B}\|\varphi_{xx}\|_\infty^2 \,\mathrm{d}s$$

$$= 2(1-\delta)\int_0^t \lambda(t_0 + s)\,\mathrm{d}s + \frac{9\delta}{2\varepsilon_B}\int_0^t \|\varphi_{xx}\|_\infty^2\,\mathrm{d}s$$

$$\leq 2(1-\delta)\int_0^t \lambda^*(t_0 + s)\,\mathrm{d}s + \frac{9\delta}{2\varepsilon_B}\int_0^t \|\varphi_{xx}\|_\infty^2\,\mathrm{d}s$$

$$= 2(1-\delta)\int_0^t (1 - \tfrac{s}{h})\lambda(t_0) + \tfrac{s}{h}\lambda(t_1)\,\mathrm{d}s + \frac{9\delta}{2\varepsilon_B}\int_0^t \|\varphi_{xx}\|_\infty^2\,\mathrm{d}s$$

$$= 2(1-\delta)\Big((t - \tfrac{t^2}{2h})\lambda(t_0) + \tfrac{t^2}{2h}\lambda(t_1)\Big) + \frac{9\delta}{2\varepsilon_B}\int_0^t \|\varphi_{xx}\|_\infty^2\,\mathrm{d}s,$$

where we reuse the bound from (5.4) for $\int_0^t \|\varphi_{xx}\|_\infty^2\,\mathrm{d}s$ to obtain

$$A(t) \leq 2(1-\delta)\Big((t - \frac{t^2}{2h})\lambda(t_0) + \frac{t^2}{2h}\lambda(t_1)\Big)$$

$$+ \frac{9\delta}{2\varepsilon_B}\Big((t - \frac{t^2}{h} + \frac{t^3}{3h^2})\|\varphi_{xx}(t_0)\|_\infty^2$$

$$+ 2\Big(\frac{t^2}{2h} - \frac{t^3}{3h^2}\Big)\|\varphi_{xx}(t_0)\|_\infty\|\varphi_{xx}(t_1)\|_\infty$$

$$+ \frac{t^3}{3h^2}\|\varphi_{xx}(t_1)\|_\infty^2\Big).$$

The special case for $t = h$, which we will need later, is given by the reduced term

$$A(h) \leq h(1 - \delta)\Big(\lambda(t_0) + \lambda(t_1)\Big) + \frac{9\delta}{2\varepsilon_B}\Big(\frac{h}{3}\|\varphi_{xx}(t_0)\|_\infty^2$$
$$+ \frac{h}{3}\|\varphi_{xx}(t_0)\|_\infty\|\varphi_{xx}(t_1)\|_\infty + \frac{h}{3}\|\varphi_{xx}(t_1)\|_\infty^2\Big).$$

Therefore, the first difference can be treated similar to (5.5) and we obtain

$$A(h) - A(s) \leq 2(1 - \delta)\Big(((h - s) - \frac{h^2 - s^2}{2h})\lambda(t_0) + \frac{h^2 - s^2}{2h}\lambda(t_1)\Big)$$
$$+ \frac{9\delta}{2\varepsilon_B}\Big(\Big(h - s - \frac{h^2 + s^2}{h} + \frac{h^3 - s^3}{3h^2}\Big)\|\varphi_{xx}(t_0)\|_\infty^2$$
$$+ 2\Big(\frac{h^2 - s^2}{2h} - \frac{h^3 + s^3}{3h^2}\Big)\|\varphi_{xx}(t_0)\|_\infty\|\varphi_{xx}(t_1)\|_\infty$$
$$+ \frac{h^3 - s^3}{3h^2}\|\varphi_{xx}(t_1)\|_\infty^2\Big).$$

We can now use these results to bound T as

$$e^{A(h)}z(t_0) \leq \exp\Big\{h(1 - \delta)\Big(\lambda(t_0) + \lambda(t_1)\Big) + \frac{9\delta}{2\varepsilon_B}\Big(\frac{h}{3}\|\varphi_{xx}(t_0)\|_\infty^2$$
$$+ \frac{h}{3}\|\varphi_{xx}(t_0)\|_\infty\|\varphi_{xx}(t_1)\|_\infty + \frac{h}{3}\|\varphi_{xx}(t_1)\|_\infty^2\Big)\Big\}z(t_0)$$

and

$$\int_0^h e^{4(A(h)-A(s))}\|\operatorname{Res}\|_{-1}^2 \, \mathrm{d}s$$
$$\leq \exp\Big\{8(1 - \delta)\kappa + \frac{18\delta}{\varepsilon_B}\Big(\frac{h}{3}\|\varphi_{xx}(t_0)\|_\infty^2$$
$$+ \frac{h}{3}\|\varphi_{xx}(t_0)\|_\infty\|\varphi_{xx}(t_1)\|_\infty + \frac{h}{3}\|\varphi_{xx}(t_1)\|_\infty^2\Big)\Big\}$$
$$\times \int_0^h C_0 + sC_1 + s^2C_2 + s^3C_3 + s^4C_4 \, \mathrm{d}s$$

where

$$\kappa = \begin{cases} h \cdot \max(\lambda(t_0), \lambda(t_1)) & \max(\lambda(t_0), \lambda(t_1)) \geq 0 \\ 0 & \text{else} \end{cases}.$$

Here, we did the same as before in (5.7). The only difference is, that this time we do not know the sign of $\lambda(t)$, which leads to κ and its different cases.

Again, we lose quite a bit if it is negative, but in order to move the exponential term out of the integral we have to do this. As the previous two estimates deliver the upper bound to T, the only term left is $\int_0^h \exp(4(A(s) - A(h)))\, ds$, for which we first need a bound for $A(s) - A(h)$.

$$A(s) - A(h) = (2\delta - 2) \int_s^h \lambda(t_0 + \tau)\, d\tau - \frac{9\delta}{2\varepsilon_B} \int_s^h \|\varphi_{xx}\|_\infty^2\, d\tau$$
$$\leq (2\delta - 2) \ldots ?$$

and this is where we get stuck. We can simply discard the second term, but there is no feasible way to obtain a rigorous and computable lower bound on $\int_s^h \lambda(t_0 + \tau)\, d\tau$.

How we can fix it

What we can do, however, is to replace $\lambda(t)$ in our very first ODE (5.9) with either the linear upper bound of $(1 - \frac{t}{h})\lambda(t_0) + \frac{t}{h}\lambda(t_1)$ or even more simplified with just $\max(\lambda(t_0), \lambda(t_1))$. This gives us a very clear time dependence, respectively removes it completely. And by doing it in the underlying ODE itself, we have no problem justifying it. This is kind of astonishing, as we are able to do this intuitive and crucial estimate only at the beginning, and not after the application of our ODE estimate (Theorem 2.7 respectively 3.3).

We start again with the ODE (5.9) and apply the estimate $\lambda(t) \leq \max(\lambda(t_0), \lambda(t_1)) =: \bar{\lambda}$

$$\partial_t \|d_x\|^2 \leq \frac{7^7 \cdot 2}{4^8 (\delta \varepsilon_C)^7} \|d_x\|^{10} + 2\left((1 - \delta)\bar{\lambda} + \frac{9\delta}{4\varepsilon_B} \|\varphi_{xx}\|_\infty^2\right) \|d_x\|^2 \qquad (5.10)$$
$$+ \frac{1}{2\delta \varepsilon_D} \|\operatorname{Res}\|_{-1}^2.$$

We can basically repeat everything from the first try now, after we apply Method 3 (Theorem 3.3)

$$\|d_x\|^2 \leq T \times \left(1 - 4 \times T^4 \frac{7^7 \cdot 2}{4^8 (\delta \varepsilon_C)^7} \int_0^h \exp\left(4(A(s) - A(h))\right)\, ds\right)^{-1/4}$$

where

$$T := e^{A(h)} z(0) + \frac{1}{2\delta \varepsilon_D} \int_0^h e^{A(h) - A(s)} \|\operatorname{Res}(s)\|_{-1}^2\, ds$$

and

$$A(t) = 2 \int_0^t (1 - \delta)\bar{\lambda} + \frac{9\delta}{4\varepsilon_B} \|\varphi_{xx}(s)\|_\infty^2\, ds.$$

For $A(t)$ we have

$$A(t) \leq 2(1-\delta)t\bar{\lambda} + \frac{9\delta}{2\varepsilon_B}\Big(\big(t - \frac{t^2}{h} + \frac{t^3}{3h^2}\big)\|\varphi_{xx}(t_0)\|_\infty^2$$
$$+ 2\Big(\frac{t^2}{2h} - \frac{t^3}{3h^2}\Big)\|\varphi_{xx}(t_0)\|_\infty\|\varphi_{xx}(t_1)\|_\infty + \frac{t^3}{3h^2}\|\varphi_{xx}(t_1)\|_\infty^2\Big) \tag{5.11}$$

and therefore

$$A(h) - A(s) \leq 2(1-\delta)(h-s)\bar{\lambda}$$
$$+ \frac{9\delta}{2\varepsilon_B}\Big(\big(h - s - \frac{h^2+s^2}{h} + \frac{h^3-s^3}{3h^2}\big)\|\varphi_{xx}(t_0)\|_\infty^2$$
$$+ 2\Big(\frac{h^2-s^2}{2h} - \frac{h^3+s^3}{3h^2}\Big)\|\varphi_{xx}(t_0)\|_\infty\|\varphi_{xx}(t_1)\|_\infty \tag{5.12}$$
$$+ \frac{h^3-s^3}{3h^2}\|\varphi_{xx}(t_1)\|_\infty^2\Big).$$

Hence, we can bound T using

$$e^{A(h)}z(t_0) \leq \exp\Big\{2(1-\delta)h\bar{\lambda} + \frac{9\delta}{2\varepsilon_B}\Big(\frac{h}{3}\|\varphi_{xx}(t_0)\|_\infty^2$$
$$+ \frac{h}{3}\|\varphi_{xx}(t_0)\|_\infty\|\varphi_{xx}(t_1)\|_\infty + \frac{h}{3}\|\varphi_{xx}(t_1)\|_\infty^2\Big)\Big\}z(t_0) \tag{5.13}$$

and

$$\int_0^h e^{A(h)-A(s)}\|\operatorname{Res}\|_{-1}^2 \, \mathrm{d}s$$
$$\leq \exp\Big\{2(1-\delta)h\max(0,\bar{\lambda}) + \frac{9\delta}{2\varepsilon_B}\Big(\frac{h}{3}\|\varphi_{xx}(t_0)\|_\infty^2$$
$$+ \frac{h}{3}\|\varphi_{xx}(t_0)\|_\infty\|\varphi_{xx}(t_1)\|_\infty + \frac{h}{3}\|\varphi_{xx}(t_1)\|_\infty^2\Big)\Big\} \tag{5.14}$$
$$\times \int_0^h C_0 + sC_1 + s^2C_2 + s^3C_3 + s^4C_4 \, \mathrm{d}s.$$

For the last estimate, where we got stuck before, we now have

$$A(s) - A(h) = (2\delta-2)(h-s)\bar{\lambda} - \frac{9\delta}{2\varepsilon_B}\int_s^h \|\varphi_{xx}\|_\infty^2 \, \mathrm{d}\tau$$
$$\leq (2\delta-2)(h-s)\bar{\lambda} - \frac{9\delta}{2\varepsilon_B}(h-s)\min_{t\in[t_0,t_1]}\|\varphi_{xx}\|_\infty^2 \tag{5.15}$$

and therefore

$$\int_0^h \exp\Big(4(A(s) - A(h))\Big) \, \mathrm{d}s$$

$$\leq \int_0^h \exp\left(4(2\delta - 2)(h - s)\bar{\lambda} - \frac{18\delta}{\varepsilon_B}(h - s)\min_{t \in [t_0, t_1]}\|\varphi_{xx}\|_\infty^2\right)\,\mathrm{d}s$$

$$= \int_0^h \exp\left(\underbrace{\left((8\delta - 8)\bar{\lambda} - \frac{18\delta}{\varepsilon_B}\min_{t \in [t_0, t_1]}\|\varphi_{xx}\|_\infty^2\right)}_{C}(h - s)\right)\,\mathrm{d}s$$

$$= e^{Ch}\int_0^h e^{-Cs}\,\mathrm{d}s = \frac{e^{Ch} - 1}{C} = \frac{e^{((8\delta-8)\bar{\lambda} - \frac{18\delta}{\varepsilon_B}\min_{t \in [t_0,t_1]}\|\varphi_{xx}\|_\infty^2)h} - 1}{(8\delta - 8)\bar{\lambda} - \frac{18\delta}{\varepsilon_B}\min_{t \in [t_0,t_1]}\|\varphi_{xx}\|_\infty^2}.$$

We can now summarize everything in the following theorem, which will define our Method 5.

Theorem 5.2 (Method 5)**.** *Under the setting of Theorem 3.3 and Theorem 4.1, with $t_{i+1} - t_i = h$ it holds that*

$$z(t_{i+1}) \leq T \times \left(1 - 4 \times T^4 \frac{7^7 \cdot 2}{4^8(\delta\varepsilon_C)^7}\,D\right)^{-1/4},$$

where

$$T = \exp\left\{2(1 - \delta)h\bar{\lambda} + \frac{9\delta}{2\varepsilon_B}\left(\frac{h}{3}\|\varphi_{xx}(t_0)\|_\infty^2\right.\right.$$
$$\left.+ \frac{h}{3}\|\varphi_{xx}(t_0)\|_\infty\|\varphi_{xx}(t_1)\|_\infty + \frac{h}{3}\|\varphi_{xx}(t_1)\|_\infty^2\right)\bigg\}z(t_i)$$
$$+ \frac{1}{2\delta\varepsilon_D}\exp\left\{2(1 - \delta)h\max(0, \bar{\lambda}) + \frac{9\delta}{2\varepsilon_B}\left(\frac{h}{3}\|\varphi_{xx}(t_0)\|_\infty^2\right.\right.$$
$$\left.\left.+ \frac{h}{3}\|\varphi_{xx}(t_0)\|_\infty\|\varphi_{xx}(t_1)\|_\infty + \frac{h}{3}\|\varphi_{xx}(t_1)\|_\infty^2\right)\right\}$$
$$\times \left(h\int_0^{2\pi} a(x)^2\,\mathrm{d}x + h^2\int_0^{2\pi} a(x)b(x)\,\mathrm{d}x\right.$$
$$+ \tfrac{1}{3}h^3\int_0^{2\pi} b(x)^2 + 2a(x)c(x)\,\mathrm{d}x + \tfrac{1}{2}h^4\int_0^{2\pi} b(x)c(x)\,\mathrm{d}x$$
$$\left.+ \tfrac{1}{5}h^5\int_0^{2\pi} c(x)^2\,\mathrm{d}x\right),$$

$$D = \frac{e^{((8\delta-8)\bar{\lambda} - \frac{18\delta}{\varepsilon_B}\min_{t \in [t_0,t_1]}\|\varphi_{xx}\|_\infty^2)h} - 1}{(8\delta - 8)\bar{\lambda} - \frac{18\delta}{\varepsilon_B}\min_{t \in [t_0,t_1]}\|\varphi_{xx}\|_\infty^2}$$

and

$$a(x) = \tfrac{1}{h}(\Phi(x, t_{i+1}) - \Phi(x, t_i)) + \partial_x^3\varphi(x, t_i) + \partial_x(\partial_x\varphi(x, t_i))^2,$$
$$b(x) = \tfrac{1}{h}[\partial_x^3(\varphi(x, t_{i+1}) - \varphi(x, t_i)) + 2\partial_x(\partial_x(\varphi(x, t_{i+1}) - \varphi(x, t_i))\partial_x\varphi(x, t_i))],$$
$$c(x) = \tfrac{1}{h^2}\partial_x(\partial_x(\varphi(x, t_{i+1}) - \varphi(x, t_i)))^2,$$
$$\Phi(x) = \int \varphi(x)\,\mathrm{d}x.$$

Proof. Apply the previous estimates (5.2), (5.13), (5.14), (5.15) and Theorem 3.3 to the differential inequality (5.10).   □

Bounding the coefficients early

If we follow this workaround one step further, we could also replace all coefficients of the ODE (5.9) by their supremum and not only the eigenvalue. This would be a possibility to reduce the number of necessary calculations if this turns out to be a problem for interval arithmetic. Also the error that we introduce by these estimates should not be too bad as we already have very small step sizes. We make use of not so gentle estimates for Theorems 5.1 and 5.2 anyway, which would be unnecessary then.

CHAPTER 6

Simulations

We will now use Method 4 (Theorem 5.1) and Method 5 (Theorem 5.2) to test our approach and show how good the improvement by the eigenvalue estimate (Theorem 4.1) is. One final time we want to point out that for a fully rigorous proof we would still have to implement interval arithmetic.

The following figures are similar to those of Chapter 3. Plots denoted by "Method X" will show the respective bound, the "Smallness Method X" plots will show the $\mathcal{H}^1$-norm of the approximation φ surrounded by the gray area in which the smooth solution lies (the borders are given by the respective method). The red dotted line in these plots represents the threshold for the smallness criterion. The only new plot we introduce is the "Eigenvalue Estimate" plot and its zoomed in version. In this plot we combine all relevant information about the eigenvalue estimate from Chapter 4. The red dotted line represents the worst case estimate $-\frac{3}{4} + 9\|\varphi_{xx}\|_\infty^2$, the red dot/dashed line is the finite dimensional maximum eigenvalue λ_n, the solid blue line is the rigorous eigenvalue estimate from Theorem 4.1 and the blue dot/dashed line is the maximum number of modes needed (for each time step, but as we do not adjust our number of modes throughout the simulation, we have to ensure that N, the used number of modes, is larger than the maximum of this graph). In the original plot over the whole simulated time one usually only recognizes a decline to ≈ 0 of all values. Therefore, we provide the zoomed version to better identify the ratios between these quantities. Please note, that the worst case estimate and the eigenvalue estimate become negative for t large enough, but due to the scaling of the y-axis this is not easily recognizable.

For a parameter set of $N = 128$ and $h = 10^{-6}$, the mean execution time to reach T^* was about 10 hours ,for the initial values we used. (see Chapter B for details about the hardware). If necessary (due to N^*) we raised the number of modes which resulted in a significantly longer execution time (for reaching T^*). It is obvious that neither a further decrease of the time step-size h nor a further increase in the number of modes N would be reasonable on the used hardware. Therefore, we also stopped the execution of these extremely long simulations when smallness was reached, which would be sufficient to prove the existence of a of a global smooth and unique solution. Experience also showed that if the time criteria is reached, the bound also reaches the smallness criteria.

As it would be easy to flood this thesis with simulations and pictures of them, we will limit ourselves to some exemplary cases that we will show in detail. A recap of these simulations can be found in Table 6.1.

In Figure 6.1 we test the standard initial value $u_0 = \sin(x)$ from Chapter 3. Both methods are able to reach the smallness and time criterion comfortably. But even in this case, the bound given by Method 5 is by a factor of 10^{-5} smaller than the one given by Method 4. Note, that this initial value is already far from the set that is covered by analytic methods (i.e. smallness).

The following Figures 6.2 and 6.3 show the influence of N and h on each method. We can see, that an increased number of modes reduces the rigorous eigenvalue bound (as expected) and therefore improves Method 5. Method 4, on the other hand, does not benefit from a larger number of modes as long as a smaller N is sufficient to approximate the solution good enough (i.e. the residual is small). In regard of the time step size, both methods profit of a smaller h, but Method 5 with its smaller bound has more "room" to increase h and is therefore able to speed up the simulation (Figure 6.1 needs approx. 10 hours and Figure 6.3 approx. 3 minutes). The increase of h by a factor of 10 leads to an increase of factor 10^2 in the bound for both methods.

Figure 6.4 and Figure 6.5 now raise the difficulty as they roughly double the $\mathcal{H}^1$-norm of the initial value compared to Figure 6.1 The simulations are quite similar, as in both cases Method 4 fails relatively fast, whereas Method 5 succeeds in the time and smallness criteria. The reason for this can be seen in the eigenvalue estimate plots, as the worst case estimate used in Method 4 is a lot larger than the rigorous eigenvalue bound used in Method 5.

Figure 6.6 is a special example where we had to raise the number of modes in order to ensure $N^* \leq N$ (to validate the eigenvalue estimate). Here Method 4 is without any chance and fails more or less right at the start (Note, the edgy appearance of the graph in Figure 6.6a is the result of the fact that for plotting we only save data every $\Delta t = 0.001$). Please note that this simulation was stopped after all methods either failed or reached the

u_0	$\|u_0\|$	$\|u_x(0)\|$	T^*	N^*	N	h	Smallness M4	Smallness M5	Time M4	Time M5	Figure
$\sin(x)$	1.77	1.77	3.54	44	128	10^{-5}	1.38	1.17	✓	✓	6.1
$\sin(x)$	1.77	1.77	3.54	44	64	10^{-5}	1.38	1.17	✓	✓	6.2
$\sin(x)$	1.77	1.77	3.54	44	64	10^{-4}	–	1.17	0.59	✓	6.3
$\frac{3}{2}\cos(x) - \frac{1}{2}\sin(2x) + \frac{1}{3}\cos(3x)$	2.86	3.65	5.7	201	256	10^{-5}	–	–	0.03	0.6	–
$\frac{3}{2}\cos(x) - \frac{1}{2}\sin(2x) + \frac{1}{3}\cos(3x)$	2.86	3.65	5.7	201	256	10^{-6}	–	1.48	0.04	✓	6.4
$2\sin(x)$	3.54	3.54	7.1	142	256	10^{-5}	–	–	0.07	0.3	–
$2\sin(x)$	3.54	3.54	7.1	142	256	10^{-6}	–	1.75	0.08	✓	6.5
$\sin(2x) + \cos(2x)$	2.51	5.01	5.01	346	512	10^{-6}	–	0.13	0.01	*	6.6
$2\cos(x) + \sin(2x)$	3.96	5.01	7.93	395	512	10^{-6}	–	–	0.01	0.03	6.7
$1.5\sin(x) + \sin(2x)$	3.20	4.43	6.39	327	512	10^{-6}	–	–	0.01	0.37	6.8

Table 6.1: Summary of all simulations of Chapter 6. u_0 is the initial value, $\|u_0\|$ the L^2-norm (important for the time criterion) and $\|u_x(0)\|$ the $\mathcal{H}^1$-norm (important for the smallness criterion). By T^* we denote the time necessary for the time criterion and by N^* the number of modes necessary for the rigorous eigenvalue estimate. N is the number of Fourier modes used ($N \geq N^*$ has to hold for Method 5) and h the step size in time. If a method succeeds in the smallness criterion, we give the respective time, and for the time criterion, we either note the time of the blowup or give a "✓" if it succeeded (the $*$ indicates that the simulation was stopped after reaching the smallness threshold). Please note that all values are rounded to two decimals.

smallness threshold (to avoid several days of execution time just to reach T^*).

Figure 6.7 and Figure 6.8 show initial values of comparable $\mathcal{H}^1$-norm to the previous simulation in Figure 6.6 (in Figure 6.8 even smaller). But to some surprise also Method 5 fails here and it is not even close. This tells us, that although the $\mathcal{H}^1$-norm of the initial value is the quantity that controls the analytic result, it is not the factor that is relevant for our methods.

Throughout all simulations it is easy to see the huge difference between the worst case estimate and the rigorous eigenvalue bound. Therefore it is not a big surprise that Method 5 is superior to Method 4 in all simulations.

Observations and Conclusion

From the previously presented simulations and the many more that were created but had no place in this thesis, we can come to the following observations and conclusions:

- Most important, the methods work. They are not just valid as analytic estimates but also actually work when carried out in simulations. Although we did not use interval arithmetic, the "ease" with which

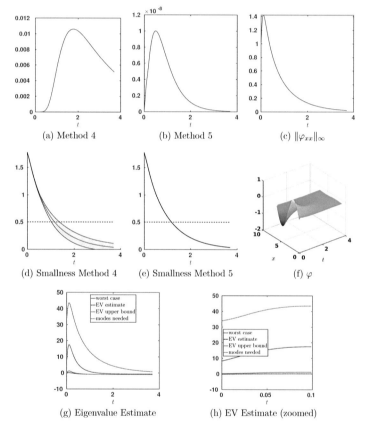

(a) Method 4 (b) Method 5 (c) $\|\varphi_{xx}\|_\infty$

(d) Smallness Method 4 (e) Smallness Method 5 (f) φ

(g) Eigenvalue Estimate (h) EV Estimate (zoomed)

Figure 6.1: $u_0 = \sin(x)$, $N = 128$ and $h = 10^{-5}$. N is larger than the maximum for modes needed, so that Method 5 is valid. Both methods comfortably succeed in both criteria.

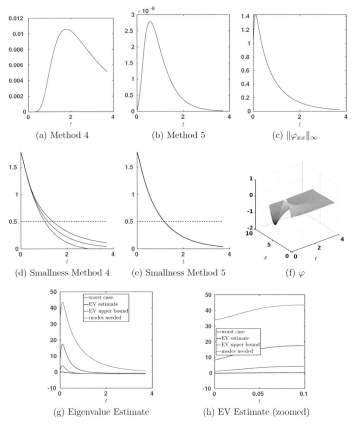

(a) Method 4   (b) Method 5   (c) $\|\varphi_{xx}\|_\infty$

(d) Smallness Method 4   (e) Smallness Method 5   (f) φ

(g) Eigenvalue Estimate   (h) EV Estimate (zoomed)

Figure 6.2: $u_0 = \sin(x)$, $N = 64$ and $h = 10^{-5}$. N is larger than the maximum for modes needed, so that Method 5 is valid. Both methods succeed in both criteria, but the bound for Method 4 is close to the region where the blowup is immediate.

58

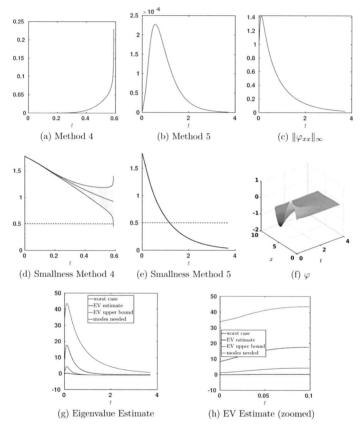

(a) Method 4   (b) Method 5   (c) $\|\varphi_{xx}\|_\infty$

(d) Smallness Method 4   (e) Smallness Method 5   (f) φ

(g) Eigenvalue Estimate   (h) EV Estimate (zoomed)

Figure 6.3: $u_0 = \sin(x)$, $N = 64$ and $h = 10^{-4}$. N is larger than the maximum for modes needed, so that Method 5 is valid. In this case Method 4 fails whereas Method 5 succeeds in both criteria and is still "comfortable" as the bound is of order 10^{-6}.

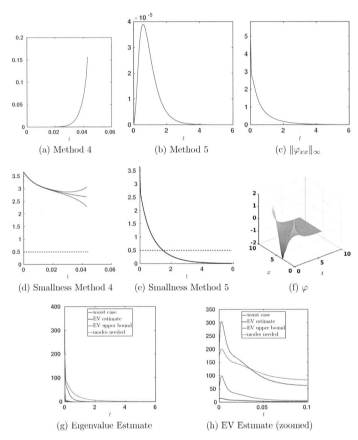

Figure 6.4: $u_0 = \frac{3}{2}\cos(x) - \frac{1}{2}\sin(2x) + \frac{1}{3}\cos(3x)$, $N = 256$ and $h = 10^{-6}$. N is larger than the maximum for modes needed, so that Method 5 is valid. Method 4 fails relatively fast whereas Method 5 succeeds in both the smallness and time criterion.

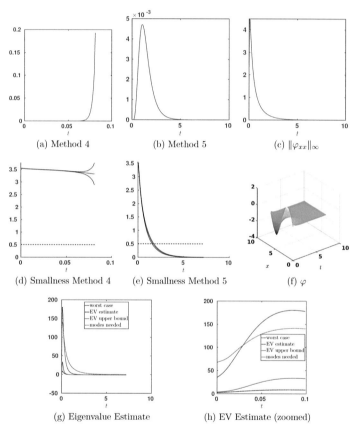

Figure 6.5: $u_0 = 2\sin(x)$, $N = 256$ and $h = 10^{-6}$. N is larger than the maximum for modes needed, so that Method 5 is valid. Method 4 fails relatively fast whereas Method 5 succeeds in both the smallness and time criterion.

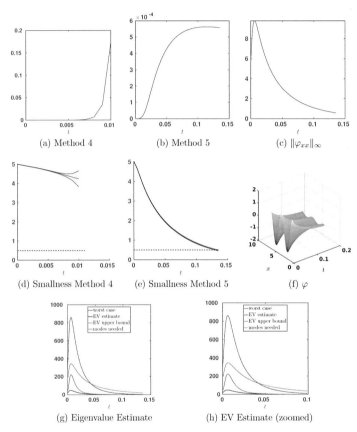

(a) Method 4   (b) Method 5   (c) $\|\varphi_{xx}\|_\infty$

(d) Smallness Method 4   (e) Smallness Method 5   (f) φ

(g) Eigenvalue Estimate   (h) EV Estimate (zoomed)

Figure 6.6. $u_0 = \sin(2x) + \cos(2x)$, $N = 512$ and $h = 10^{-6}$. N is larger than the maximum for modes needed, so that Method 5 is valid. Method 4 fails extremely fast whereas Method 5 succeeds in the smallness criterion. The simulation was stopped after Method 5 reached the smallness criterion because it would take another two days for the simulation to reach $T*$ (due to the necessary increase of N). Although, the declining bound of Method 5 and experience indicate that T^* would have been reached.

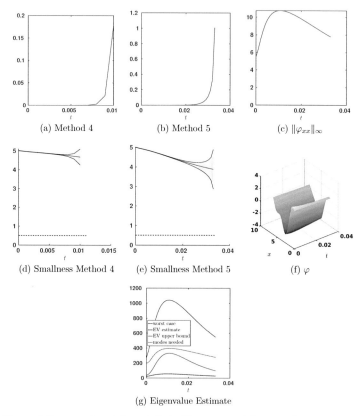

(a) Method 4   (b) Method 5   (c) $\|\varphi_{xx}\|_\infty$

(d) Smallness Method 4   (e) Smallness Method 5   (f) φ

(g) Eigenvalue Estimate

Figure 6.7: $u_0 = 2\cos(x) + \sin(2x)$, $N = 512$ and $h = 10^{-6}$. N is larger than the maximum for modes needed, so that Method 5 is valid, but both methods fail very fast.

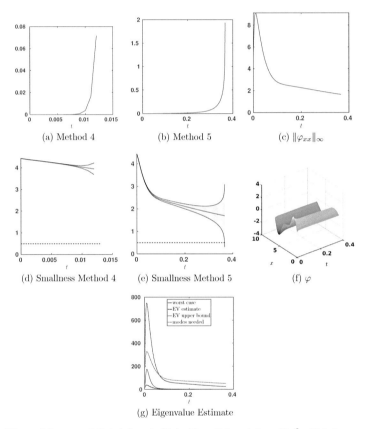

(a) Method 4   (b) Method 5   (c) $\|\varphi_{xx}\|_\infty$

(d) Smallness Method 4   (e) Smallness Method 5   (f) φ

(g) Eigenvalue Estimate

Figure 6.8: $u_0 = 1.5\sin(x) + \sin(2x)$, $N = 512$ and $h = 10^{-6}$. N is larger than the maximum for modes needed, so that Method 5 is valid, but both methods fail very fast.

both methods, but particularly Method 5, "solved" initial values like $u_0 = \sin(x)$ gives reason to be optimistic that this will also hold using interval arithmetic.

- The eigenvalue estimate is a huge improvement. All simulations show a clear advantage of Method 5 over Method 4 by a factor between 10^{-3} and 10^{-5} in the bound and between 3 and 4 for the eigenvalue estimate vs. worst case estimate.

- For all of our simulations it holds that if T^* is reached, smallness will also be reached. Although there is no proof that one criterion implies the other (for our bounds!), we have some heuristic arguments for each implication.
  For the first direction "smallness" $\Rightarrow$ "time", a good explanation is, that if we take a closer look at the simulations, we see that the time when a bound decreases and the time when it fulfills the smallness criterion are roughly the same. Adding the observation that a bound does not increase again after it peaked once, this also leads to the conclusion that the bound stays finite until T^*. On the other hand, it seems possible to construct an initial value where the blowup occurs just right in between both events. Therefore, although this should be true for most of the simulations, this does not look like a strict implication. The other direction, "time" $\Rightarrow$ "smallness" is a lot easier to explain and also something where we can be optimistic that it in fact is a true implication. The approximation tends very fast to 0, and the error bound from the method can not get too large, especially not as large as 0.5 where the smallness threshold is, or it would have a blowup anyway. But this can not happen because it reached T^*. Therefore, the bound has to stay small and it is dragged by the approximation below the threshold.

- Other than expected by the analytic result, our methods are not (directly) sensitive to the $\mathcal{H}^1$-norm (see Figures 6.7, 6.8 and 6.6). The driving quantity for our methods is $\|u_{xx}\|_\infty$.

- The effect of more modes or/and a smaller step size could not be investigated in more depth due to hardware limitations on one side and the necessity of a minimal number of modes given by the eigenvalue estimate on the other. The only study in this respect was done with the initial value of $u_0 = \sin(x)$ (see Figures 6.1, 6.2 and 6.3). From these simulations we can see, that an increased number of modes reduces the rigorous eigenvalue bound directly (as expected by the theorem) and therefore improves Method 5. The improvement by a smaller residual is shared by both methods and is only recognizable if N was too small compared to the frequencies of the solution. In regard of the time step size, both methods profit of a smaller h, but Method 5 with its smaller bound has more "room" to increase h and by this, to speed up the simulation (The time for Figure 6.1 is approx. 1 hour and for Figure 6.3 approx. 3 minutes). The increase of h by a factor of 10 led to an increase of factor 10^2 in the bound for both methods.

- Our methods and simulations are very fragile with respect to high frequencies (and high in this case starts at around 3 if not dampened). Due to the important role of $\|\varphi_{xx}\|_\infty$ for our estimates, these lead to a very fast blowup of the methods.

  Also our semi-implicit Euler method (see Chapter B) is unstable if the frequency or amplitudes of the initial value are too large. The numerical problem could be avoided by using a different scheme, e.g. an exponential Euler method.

  For the methods itself one could circumvent this problem in some special cases by using the scaling property (Theorem A.10) of the surface growth equation (1.1) to represent solutions for initial values with high frequencies by those with lower frequencies (see Section A.3).

- Possible new approaches to improve these results might have to get rid of the exponential growth that we introduce with our Gronwall methods (which would hopefully also remove the strong sensitivity concerning $\|\varphi_{xx}\|_\infty$). Maybe fixed point methods in the way they are used in e.g. [BL15] are a possibility to achieve this.

APPENDIX A

Basic Tools

This chapter is a brief collection of all the estimates from basic calculus, ODE and PDE theory, that we use throughout the thesis. These can be found in e.g. [Eva10] (see also its collection of inequalities in Appendix B) Additionally, we will give a proof for the frequently used L^2-energy estimate for the surface growth equation.

A.1 General Inequalities

The first is Young's inequality which is one of the basic tools of calculus and can be used to prove Hölder's inequality for example.

Theorem A.1. *(Young's inequality)*
If $a, b \geq 0$ and $p, q > 0$ such that $\frac{1}{p} + \frac{1}{q} = 1$, then

$$ab \leq \frac{a^p}{p} + \frac{b^q}{q}$$

If we set $a := (\varepsilon p)^{1/p} a$ and $b := (\varepsilon p)^{-1/p} b$ we get the scaled version of Young's inequality, also called "Young's inequality with ε", which is a more generalized version that allows us to weight the factors differently.

Corollary A.2. *(Young's inequality with ε)*
If $a, b \geq 0$, $\varepsilon > 0$ and $p, q > 0$ such that $\frac{1}{p} + \frac{1}{q} = 1$ then

$$ab \leq \varepsilon a^p + \frac{(p\varepsilon)^{1-q}}{q} b^q$$

Next is the Poincaré inequality, which is one of the fundamental inequalities of Sobolev spaces.

Theorem A.3. *(Poincaré inequality)*
Assume $1 \leq p \leq \infty$ and that $\Omega \subset \mathbb{R}^n$ is a bounded connected open subset with Lipschitz boundary.
Then for every $u \in W^{1,p}(\Omega)$ there exists a constant C, only depending on Ω and p, such that

$$\left\| u - \frac{1}{|\Omega|} \int_\Omega u(x) \, \mathrm{d}x \right\|_{L^p(\Omega)} \leq C \left\| \nabla u \right\|_{L^p(\Omega)}.$$

Proof. For a proof see [Eva10] Chapter 5.6 Theorem 3. □

In the special setting with $\Omega = [0, 2\pi]$ and $u \in C^1$ as a 2π-periodic function, which is the setting for our surface growth equation, the Poincaré inequality is also known as Wirtinger's inequality. Note that the Poincaré constant $C = 1$ is sharp in this case.

Corollary A.4. *(Wirtinger's inequality I)*
Let $u \in C^1([0, 2\pi], \mathbb{R})$ be a 2π-periodic function such that

$$\int_{[0,2\pi]} u(x) \, \mathrm{d}x = 0.$$

Then

$$\left\| u' \right\|_{L^2([0,2\pi])} \geq \left\| u \right\|_{L^2([0,2\pi])}.$$

The following version of Agmon's inequality is a special case adapted to our settings. For the general case, the constant on the right hand side is not equal to 1.

Theorem A.5. *(Agmon's inequality)*
Let $u \in H^1([0, 2\pi], \mathbb{R}) \cap C^0([0, 2\pi], \mathbb{R})$ be a 2π-periodic function such that

$$\int_0^{2\pi} u(x) \, \mathrm{d}x = 0.$$

Then

$$\left\| u \right\|_{L^\infty([0,2\pi])}^2 \leq \left\| u \right\|_{L^2([0,2\pi])} \left\| \partial_x u \right\|_{L^2([0,2\pi])}.$$

Proof. As $\int_0^{2\pi} u(x) \, \mathrm{d}x = 0$ there exists a $\xi \in [0, 2\pi]$ such that $u(\xi) = 0$. From

$$u^2(x) - u^2(\xi) = \int_\xi^x \partial_x(u^2) \, \mathrm{d}y = 2 \int_\xi^x \partial_x u \cdot u \, \mathrm{d}y \qquad \forall x \in [0, 2\pi]$$

it follows that

$$u^2(x) = \underbrace{u^2(\xi)}_{=0} + 2\int_\xi^x \partial_x u \cdot u \; \mathrm{d}y \le 2\int_\xi^x |\partial_x u \cdot u| \; \mathrm{d}y,$$

and also, because u is 2π-periodic,

$$u^2(x) = \underbrace{u^2(\xi + 2\pi)}_{=0} - 2\int_x^{\xi+2\pi} \partial_x u \cdot u \; \mathrm{d}y \le 2\int_x^{\xi+2\pi} |\partial_x u \cdot u| \; \mathrm{d}y.$$

Now add up both inequalities and we have

$$u^2(x) \le \int_\xi^{\xi+2\pi} |\partial_x u \cdot u| \; \mathrm{d}y = \int_0^{2\pi} |\partial_x u \cdot u| \; \mathrm{d}y,$$

again because of the periodicity. Now apply Cauchy-Schwarz to finish the proof:

$$\|u\|_{L^\infty([0,2\pi])}^2 \le \|u\|_{L^2([0,2\pi])} \|\partial_x u\|_{L^2([0,2\pi])}.$$

$\square$

In addition to the Poincaré inequality, the Interpolation inequality gives us another tool to compare the L^2-norms of a function and its (weak) derivatives.

Theorem A.6. *(Interpolation inequality)*
Assume $u \in H^2([0,2\pi], \mathbb{R}) \cap H_0^1([0,2\pi], \mathbb{R})$ then

$$\|\partial_x u\|_{L^2([0,2\pi])} \le \|u\|_{L^2([0,2\pi])}^{\frac{1}{2}} \left\|\partial_x^2 u\right\|_{L^2([0,2\pi])}^{\frac{1}{2}}.$$

Proof.

$$\|\partial_x u\|_{L^2([0,2\pi])} = \left(\int_0^{2\pi} \partial_x u \cdot \partial_x u \; \mathrm{d}x\right)^{\frac{1}{2}}$$

$$= \left(\underbrace{[\partial_x u \cdot u]_0^{2\pi}}_{=0} - \int_0^{2\pi} u \cdot \partial_x^2 u \; \mathrm{d}x\right)^{\frac{1}{2}} \qquad \text{int. by parts}$$

$$= \left(\left|\int_0^{2\pi} u \cdot \partial_x^2 u \; \mathrm{d}x\right|\right)^{\frac{1}{2}}$$

$$\le \|u\|_{L^2([0,2\pi])}^{\frac{1}{2}} \left\|\partial_x^2 u\right\|_{L^2([0,2\pi])}^{\frac{1}{2}} \qquad \text{by C.S. ineq.}$$

$\square$

Gronwall's inequality is an important piece of the uniqueness theory of ODEs.

Theorem A.7. *(Gronwall inequality)*
Assume $b(t) > 0$, $a(t) \in \mathbb{R}$ *and* $u \in L^1([0,T],\mathbb{R})$ *such that*

$$u(t) \leq a(t) + \int_0^t b(s) \cdot u(s) \, \mathrm{d}s \qquad \forall t \in [0,T].$$

Then

$$u(t) \leq a(t) + \int_0^t a(s) \cdot b(s) \cdot \exp\left(\int_s^t b(\tau) \, \mathrm{d}\tau\right) \mathrm{d}s \qquad \forall t \in [0,T].$$

Proof. For a proof see [Eva10] Appendix B. □

Using Gronwall's inequality, one can show the following comparison principle for ODEs.

Theorem A.8. *(Comparison Principle)*
Assume $y, x \in C^1([0,T],\mathbb{R})$ *and* $f, D_2 f \in C^0([0,T] \times \mathbb{R}, \mathbb{R})$ *such that*

$$x'(t) \leq f(t, x(t)) \qquad \forall t \in [0,T]$$
$$y'(t) = f(t, y(t)) \qquad \forall t \in [0,T]$$
$$x(0) \leq y(0).$$

Then

$$x(t) \leq y(t) \qquad \forall t \in [0,T].$$

Proof. Set $h = x - y$ and assume that there is a time $0 < t^* < T$ such that $h(t^*) > 0$. Due to the continuity of x and y and hence h, there exists a time $t_0 \in [0, t^*)$ such that $h(t_0) = 0$ and $h(t) > 0$ for $t \in (t_0, t^*]$. For $t \in [t_0, t^*]$ we now have that

$$h'(t) \leq f(t, x(t)) - f(t, y(t)) \leq L|x(t) - y(t)| = Lh(t).$$

If we now apply Gronwall's inequality it follows that

$$h(t^*) \leq h(t_0)e^{L(t^* - t_0)} = 0$$

which is a contradiction to our initial assumption. □

A.2 L2 Estimate of the SFG Equation

The following estimate, or better equality, is very important for analytic results to our surface growth equation (1.1) as it shows that the L^2-scalar product of the nonlinearity with a solution u vanishes. This result and similar calculations are repeatedly used in many estimates.

Theorem A.9 (L^2-estimate). *Let u be a solution to our surface growth equation (1.1) according to Theorem 2.1.*

Then it holds that

$$\frac{1}{2}\partial_t\|u\|^2 = -\|u_{xx}\|^2,$$

because the scalar product of the nonlinearity with the solution u vanishes, i.e. $\langle (u_x{}^2)_{xx}, u \rangle = 0$.

Proof. Due to the regularity of our solution u, we have

$$\frac{1}{2}\partial_t\|u\|^2 = \langle u_t, u \rangle = -\langle u_{xxxx} + (u_x{}^2)_{xx}, u \rangle$$
$$= -\langle u_{xxxx}, u \rangle - \langle (u_x{}^2)_{xx}, u \rangle$$
$$= -\langle u_{xx}, u_{xx} \rangle - \langle (u_x{}^2)_{xx}, u \rangle$$
$$= -\|u_{xx}\|^2 - \langle (u_x{}^2)_{xx}, u \rangle.$$

Now, for the second term it holds using integration by parts that

$$\langle (u_x{}^2)_{xx}, u \rangle = \langle u_x{}^2, u_{xx} \rangle,$$

and on the other hand by differentiating

$$\langle (u_x{}^2)_{xx}, u \rangle = \langle 2u_x u_{xx}, u_x \rangle = 2\langle u_x{}^2, u_{xx} \rangle.$$

Therefore,

$$\langle (u_x{}^2)_{xx}, u \rangle = 0.$$

Combining both calculations yields the equality of the theorem. □

A.3 Scaling Property of the SFG Equation

Our surface growth equation (1.1) possesses a scaling property which can be useful to show global regularity and uniqueness of solutions with an initial value of "high" frequencies by applying our methods to an initial value of "lower" frequencies.

Theorem A.10 (Scaling Property). *Let u be a solution to our surface growth equation (1.1) on $\mathbb{R} \times \mathbb{R}^+$.*
Then u_λ, defined by

$$u_\lambda(x, t) := u(\lambda x, \lambda^4 t),$$

is also a solution to the surface growth equation.

Proof. Simply plug u_λ into the surface growth equation (1.1) and apply the chain rule. $\qquad\square$

With this property, the existence of a globally regular and unique solution for the initial value $\sin(\lambda x)$ can be reduced to the question if the initial value $\sin(x)$ possesses such a solution for example.

APPENDIX B

Numerical Implementation

Here, we will take a look at some examples of the numerical implementation. All simulations in this thesis were done using a PC with an Intel® Core™ i7-6700 CPU and 16 GB RAM running Fedora 24 (64Bit) and MATLAB® version 9.0.0.341360 (R2016a). Usual simulations took about 13 to 18 hours (reaching T^*) using step-size $h = 10^{-6}$ and $N = 128$ modes.

B.1 Approximation Method

First, remember that the only restrictions to φ are certain smoothness conditions. This is important as it allows us to use every numerical method we like. We do not have to justify the used numerical method as long as our bounds stay finite. Of course we expect better results for φ close to an actual solution, so plain guessing of φ, although valid, is not encouraged.

To compute our arbitrary approximation φ, we use a spectral Galerkin method to convert the PDE to a system of ODEs. The basis of eigenfunctions is in our case the standard Fourier basis $e_k = \frac{1}{\sqrt{2\pi}}\exp(ikx)$. Note, that this allows us as a welcome side effect to compute quantities like L^2 scalar products and norms very efficiently.

With $u := \sum_k a_k(t)e_k$ our surface growth equation (1.1) turns into the following system of ODEs

$$\sum_k a_k'(t)e_k = -\sum_k (ik)^4 a_k(t)e_k - \sum_k (ik)^2 \underbrace{\left(\sum_{s+l=k} (is)a_s(t) \times (il)a_l(t) \right)}_{b_k(t)} e_k$$

73

which is independent except for the nonlinearity. To solve this system, we now use a semi-implicit Euler scheme, i.e. we use time $t_{j+1} = t_j + h$ in the linear part, and t_j inside the nonlinearity (we could not solve for t_{j+1})

$$\frac{1}{h}(a_k(t_{j+1}) - a_k(t_j)) = -(ik)^4 a_k(t_{j+1}) - (ik)^2 b_k(t_j)$$

$$\Rightarrow \qquad (1 + h(k)^4)a_k(t_{j+1}) = a_k(t_j) + hk^2 b_k(t_j)$$

$$\Rightarrow \qquad\qquad a_k(t_{j+1}) = (1 + h(k)^4)^{-1}(a_k(t_j) + hk^2 b_k(t_j))$$

for all k.

```
1  % number of fourier modes, use powers of 2
2  N = 2^7;
3  % maximum time ( intervall [0,T] )
4  T = 1;
5  % time stepsize
6  h = 10^(-6);
7  % stepsize to take snapshots of solutions for plotting
8  step = 10^(-3);
9  % initial value as function handle
10 initValue = @(x) 2*sin(x);
11
12 % grid points for fft.
13 x = linspace(0,2*pi,N+1); x(N+1) = [];
14 % coeff. for derivatives in fourier space as col vector
15 k(1:N/2) = 0:1:N/2-1; k((N/2+1):N) = -N/2:1:-1; k = k.';
16 % lin. operator
17 M = 1./(ones(N,1) + h*k.^4);
18
19 % prealloc memory
20 % ensure that vector size is no float (happens with small step)
21 vecSize = round(T/step+1);
22 % approximation phi
23 phi = zeros(N,vecSize);
24
25 % initial value of the approximation
26 phi(:,1) = fft( initValue(x) );
27
28 for m=1:(vecSize-1)
29     % calculate step/h iterations.
30     for n=1:(step/h)
31         % check for new result entry (just implementation)
32         if (n==1)
33             r = 0;
34         else
35             r = 1;
36         end
37
38         % do one iteration step
39         % calculating the square directly without ifft/fft. done in square.m
40         phi(:,m+1) = M .* ( phi(:,m+r) + h*k.^2 .* ...
41             square(1i*k.*phi(:,m+r),true) );
42     end
43 end
```

Matlab Code B.1: Minimal example for the semi-implicit Euler method to compute our approximation φ.

The code snippet B.1 shows a minimal working example of this approximation scheme. The square that occurs in the nonlinearity, which we denote by b_k is calculated by the function `square` which simply implements the double sum for the product. Its second parameter decides if the solution is cut off to the initial size of the input vector (this represents the projection on our solution space H_n) as the square doubles the length of the spectrum. For faster calculation one could use the inverse transformation with `ifft`, calculate the square in real space and finally transform to Fourier space again. The problem with this approach is, that there might be interpolation involved what leads to inaccurate results of the amplitudes that belong to our solution space. Therefore, we stick to the slower variant of explicitly calculating the square of the sum.

B.2 Eigenvalue Estimate

For the second code example, we will take a look at the calculation of the eigenvalue estimate used for Method 5 (see Chapter 4).

In the beginning we had the bilinear form and its worst case estimate

$$
\begin{aligned}
\tilde{M}(d_x, d_x) &= -\langle d_{xx}, d_{xx}\rangle + 2\langle d_{xx}, (d_x\varphi_x)_{xx}\rangle \\
&= -\|d_{xxx}\|^2 + 2\int d_{xx}(d_x\varphi_x)_{xx}\,\mathrm{d}x \\
&\leq \left(-\frac{3}{4} + 9\|\varphi_{xx}\|_{L^\infty}^2\right)\|d\|_{H^1}^2.
\end{aligned}
$$

For the better eigenvalue estimate, we were looking for a (hermitian) Matrix M, such that

$$
\tilde{M}(g, g) = g^T M \bar{g},
$$

in order to compute the largest eigenvalue of M, which depends on the numerical data φ. If we had the representing matrix, we could use the fact that if a matrix $M \in \mathbb{C}^{N \times N}$ is hermitian, i.e. $M^H := \overline{M}^T = M$, there exists a unitary matrix U, i.e. $U^* = U^H = U^{-1}$, and a real diagonal matrix D such that $M = UDU^*$. And as we are interested in $\sup\{\langle x, Mx\rangle; \|x\| = 1\}$ for a hermitian matrix M, it follows that

$$
\begin{aligned}
\sup_{\|x\|=1} \langle x, Mx\rangle &= \sup_{\|x\|=1} \langle x, UDU^*x\rangle = \sup_{\|x\|=1} \langle U^*x, MU^*x\rangle = \sup_{\|y\|=1} \langle y, Dy\rangle \\
&\leq \sup_{\|y\|=1} \lambda_{\max}\langle y, y\rangle = \sup_{\|y\|=1} \lambda_{\max}\|y\|^2 = \lambda_{\max},
\end{aligned}
$$

because for $y = U^*x$ holds that $\|x\| = \|y\|$ as U is unitary.

To set up the matrix, we simply brute-force evaluate the entries in Fourier space by

$$
\begin{aligned}
M_{kl} = \tilde{M}(e_k, e_l) &= -\langle -K^2 e_k, -K^2 e_l \rangle + 2\langle iKe_k, (e_l\varphi_x)_{xx} \rangle \\
&= -\langle e_k, K^4 e_l \rangle + 2\langle iKe_k, -K^2(e_l\varphi_x) \rangle \\
&= -\langle e_k, K^4 e_l \rangle + 2\langle e_k, iK^3(e_l\varphi_x) \rangle \\
&= -\delta_{k,l}k^4 + 2\langle e_k, iK^3(e_l\varphi_x) \rangle \\
&= -\delta_{k,l}k^4 - 2ik^3\overline{(\varphi_x)}_{k-l},
\end{aligned}
$$

with $k, l = 1 \ldots N$ (we can skip 0 as the constant mode is not part of our solution space), $e_k = \frac{1}{\sqrt{2\pi}} \exp((k - 1 - N/2)ix)$ and K is the diagonal matrix representing the Fourier coefficients. Note, we do not have to normalize φ to the ONB as it is enough if e_l is already normalized with $\frac{1}{\sqrt{2\pi}}$ in the product. Though, the additional factor of N in the MATLAB fft vector of φ has to be removed. To pick $\overline{(\varphi_x)}_{k-l}$ use the corresponding Fourier vector with fftshift. In this case, the modes are ordered ascending by frequence:

$$
\text{normal mode ordering}: \begin{pmatrix} 0 \\ 1 \\ \vdots \\ N/2 - 1 \\ -N/2 \\ \vdots \\ -1 \end{pmatrix} \quad \text{after fftshift}: \begin{pmatrix} -N/2 \\ \vdots \\ 0 \\ \vdots \\ N/2 - 1 \end{pmatrix}.
$$

The matrix M is generally not hermitian/symmetric, as the product with φ acts differently on the first and second argument. The usual decomposition for complex matrices is

$$
A = \frac{1}{2}(A + A^H) + \frac{1}{2}(A - A^H) = B + iC
$$

where both, B and C, are hermitian.

For our calculation, the first part $\frac{1}{2}(A + A^H) = B$ is in fact sufficient as it holds that

$$
\begin{aligned}
\langle x, \tfrac{1}{2}(M + M^H)x \rangle &= \frac{1}{2}\Big(\langle x, Mx \rangle + \langle x, M^H x \rangle \Big) = \frac{1}{2}\Big(\langle x, Mx \rangle + \langle Mx, x \rangle \Big) \\
&= \frac{1}{2}\Big(\langle x, Mx \rangle + \overline{\langle x, Mx \rangle} \Big) = \Re\Big(\langle x, Mx \rangle \Big).
\end{aligned}
$$

At first sight, this is looking bad, as it is only the real part of the original quadratic form. But we know, that our original bilinear form is real-valued (as we only insert real functions), and therefore the corresponding bilinear form in Fourier space (restricted on vectors x representing real functions) has to be real, too.

```
1  %%%%%%%%%%%%%%%%%%%%%%%%%%%%%%%%%%%%%%%%%%%%%%%%%%%%%%%%%%%%%%%%%%%%%%%%%%
2  % [ ev, wcEst, bForm, evBound, modes] = estEV(phi,N,k)
3  % Calculates the maximal eigenvalue of the bilinear form in finite
4  % dimension, the rigorous upper bound for the eigenvalue in
5  % infinite dimension, the number of necessary modes for the estimate to
6  % hold and the former worst case estimate for comparison.
7  % arguments :
8  %     - phi : vector representing the approximation in fourier space
9  %     - N   : number of fourier modes
10 %     - k   : vector of coefficients of the derivative in fourier space
11 % return values :
12 %     - ev     : maximal eigenvalue of the bilinear form in finite space
13 %     - wcEst  : worst case estimate
14 %     - bForm  : the NxN matrix corresponding to the bilinear form
15 %     - evBound: bound for the maximal eigenvalue of the bilinear form in
16 %                whole space
17 %     - modes  : number of modes needed for rigorous bound
18 %%%%%%%%%%%%%%%%%%%%%%%%%%%%%%%%%%%%%%%%%%%%%%%%%%%%%%%%%%%%%%%%%%%%%%%%%%
19 function [ ev, wcEst, bForm, evBound, modes] = estEV( phi,N,k )
20
21 bForm = zeros(N,N);
22
23 % precalculate the needed derivatives and norms of phi
24
25 % phi_x
26 phi1 = 1i * k .* phi;
27 % Linf Norm of phi_x
28 phi1LINF = max(abs(real(ifft( phi1 ))));
29
30 % phi_xx
31 phi2 = 1i * k .* phi1;
32 % Linf Norm of phi_xx
33 phi2LINF = max(abs(real(ifft( phi2 ))));
34
35 % phi_xxx
36 phi3 = 1i * k .* phi2;
37 % Linf Norm of phi_xxx
38 phi3LINF = max(abs(real(ifft( phi3 ))));
39
40 % calculate worst case estimate
41 wcEst = -3/4 + 9 * phi2LINF^2;
42
43 % shift the fourier modes, that they are ordered -N/2 ... 0 ... N/2-1
44 % implementation is easier this way
45 % also remove the normalisation by matlab ( * 1/N )
46 phi1 = fftshift(1/N * phi1);
47
48 % calculate the matrix of the bilinear Form
49 for m = 1:N
50     for p = 1:N
51
52         % kronecker delta
53         if (m == p)
54             delta = 1;
55         else
56             delta = 0;
57         end
```

```
58
59        % check if <e_m , e_p \varphi_x> returns a phi depending value
60        % i.e. if phi at the given index is actually defined and not by
61        % definition zero (index between 1 and N)
62        if ( (m-p + N/2 + 1) >= 1 && (m-p + N/2 + 1) <= N )
63
64            bForm(m,p) = - delta * (m - N/2 - 1)^4 ...
65                - 2 * 1i * (m - N/2 - 1)^3 * conj( phi1(m-p + N/2 + 1) );
66        end
67     end
68 end
69
70 % remove row and column for the basis vector of the constant function
71 % possible as our solution space is always 0 at the constant function
72 bForm(N/2+1,:) = [];
73 bForm(:,N/2 +1) = [];
74
75 % calculate largest eigenvalue
76 bForm = 1/2 * ( bForm + bForm' );
77 ev = max(eig(bForm));
78
79 %% Second Part - calculate rigorous upper bound for max ev and the
80 %% needed number of modes.
81
82 C = (2 * phi3LINF + 6 * phi2LINF + 4 * phi1LINF);
83 % factor 2 because n from the theorem denotes the 2n-dim subspace
84 modes = 2 * sqrt(2) * C;
85
86 % only valid if N >= modes
87 n = N/2;
88 evBound = ev + 0.5 * ...
89    max(2 * C^2 * (9 * phi2LINF^2 - 2 * ev)/n^2 , ...
90    9 * phi2LINF^2 + 2 * ev - 0.5 * n^4);
91
92 end
```

Matlab Code B.2: estEv.m function that computes the eigenvalue estimate for Method 5.

List of Figures

Bibliography

[07]   *Why global regularity for Navier-Stokes is hard.* Mar. 18, 2007.
          URL: https://terrytao.wordpress.com/2007/03/18/why-
          global-regularity-for-navier-stokes-is-hard/.

[AH89]   Kenneth Appel and Wolfgang Haken. *Every planar map is four
          colorable.* Vol. 98. Contemporary Mathematics. With the collab-
          oration of J. Koch. American Mathematical Society, Providence,
          RI, 1989, pp. xvi+741. ISBN: 0-8218-5103-9.

[BL15]   Jan Bouwe van den Berg and Jean-Philippe Lessard. "Rigorous
          numerics in dynamics". In: *Notices Amer. Math. Soc.* 62.9 (2015),
          pp. 1057–1061.

[BNR15]   Dirk Blömker, Christian Nolde, and James C. Robinson. "Rig-
          orous numerical verification of uniqueness and smoothness in a
          surface growth model." In: *J. Math. Anal. Appl.* 429.1 (2015),
          pp. 311–325.

[BR09]   Dirk Blömker and Marco Romito. "Regularity and blow up in a
          surface growth model." In: *Dyn. Partial Differ. Equ.* 6.3 (2009),
          pp. 227–252.

[BR12]   Dirk Blömker and Marco Romito. "Local existence and unique-
          ness in the largest critical space for a surface growth model." In:
          NoDEA, Nonlinear Differ. Equ. Appl. 19.3 (2012), pp. 365–381.

[BR15]   Dirk Blömker and Marco Romito. "Stochastic PDEs and lack
          of regularity: a surface growth equation with noise: existence,
          uniqueness, and blow-up." In: *Jahresber. Dtsch. Math.-Ver.* 117.4
          (2015), pp. 233–286. ISSN: 0012-0456; 1869-7135/e.

[CF88]   Peter Constantin and Ciprian Foias. *Navier-Stokes Equations*.
            University of Chicago Press, 1988.

[Che+07]   Sergei I. Chernyshenko, Peter Constantin, James C. Robin-
            son, and Edriss S. Titi. "A posteriori regularity of the three-
            dimensional Navier-Stokes equations from numerical computa-
            tions." In: *J. Math. Phys.* 48.6 (2007), 065204, 15 p.

[CVG05]   R. Cuerno, L. Vázquez, and R. Gago. "Self-organized ordering of
            nanostructures produced by ion-beam sputtering". In: *Phys. Rev.
            Lett.* 94 (2005), 016102, 4 p.

[DLM07]   Sarah Day, Jean-Philippe Lessard, and Konstantin Mischaikow.
            "Validated continuation for equilibria of PDEs." In: *SIAM J. Nu-
            mer. Anal.* 45.4 (2007), pp. 1398–1424.

[DR08]   Masoumeh Dashti and James C. Robinson. "An a posteriori
            condition on the numerical approximations of the Navier-Stokes
            equations for the existence of a strong solution." In: *SIAM J.
            Numer. Anal.* 46.6 (2008), pp. 3136–3150.

[Eva10]   Lawrence C. Evans. *Partial Differential Equations: Second Edi-
            tion*. Vol. 19. AMS, 2010.

[Fef06]   Charles L. Fefferman. "Existence and smoothness of the Navier-
            Stokes equation". In: *The millennium prize problems*. Clay Math.
            Inst., Cambridge, MA, 2006, pp. 57–67.

[FV06]   Thomas Frisch and Alberto Verga. "Effect of Step Stiffness and
            Diffusion Anisotropy on the Meandering of a Growing Vicinal
            Surface". In: *Phys. Rev. Lett.* 96 (16 Apr. 2006), 166104, 4 p.

[Hen81]   Daniel Henry. *Geometric theory of semilinear parabolic equations*.
            Vol. 840. Lecture Notes in Mathematics. Springer-Verlag, Berlin-
            New York, 1981, pp. iv+348. ISBN: 3-540-10557-3.

[HKM16]   Marijn J. H. Heule, Oliver Kullmann, and Victor W. Marek.
            "Solving and Verifying the Boolean Pythagorean Triples Prob-
            lem via Cube-and-Conquer". In: *Theory and Applications of Sat-
            isfiability Testing – SAT 2016: 19th International Conference,
            Bordeaux, France, July 5-8, 2016, Proceedings*. Ed. by Nadia
            Creignou and Daniel Le Berre. Springer International Publish-
            ing, 2016, pp. 228–245. ISBN: 978-3-319-40970-2.

[KT01]   H. Koch and D. Tataru. "Well-posedness for the Navier-Stokes
            equations". In: *Adv. Math.* 157.1 (2001), pp. 22–35.

[Ler34]   Jean Leray. "Sur le mouvement d'un liquide visqueux emplissant
           l'espace." In: *Acta Math.* 63 (1934), pp. 193–248.

[Liu15]   Xuefeng Liu. "A framework of verified eigenvalue bounds for
           self-adjoint differential operators". In: *Applied Mathematics and
           Computation* (2015).

[Mai+08]   Stanislaus Maier-Paape, Ulrich Miller, Konstantin Mischaikow,
           and Thomas Wanner. "Rigorous numerics for the Cahn-Hilliard
           equation on the unit square." In: *Rev. Mat. Complut.* 21.2 (2008),
           pp. 351–426.

[MP08]   Carlo Morosi and Livio Pizzocchero. "On approximate solutions
           of semilinear evolution equations. II: Generalizations, and appli-
           cations to Navier-Stokes equations." In: *Rev. Math. Phys.* 20.6
           (2008), pp. 625–706.

[MP11]   Carlo Morosi and Livio Pizzocchero. "An H^1 setting for the
           Navier-Stokes equations: quantitative estimates." In: *Nonlin-
           ear Anal., Theory Methods Appl., Ser. A, Theory Methods* 74.6
           (2011), pp. 2398–2414.

[MP12]   Carlo Morosi and Livio Pizzocchero. "On approximate solutions
           of the incompressible Euler and Navier-Stokes equations." In:
           *Nonlinear Anal., Theory Methods Appl., Ser. A, Theory Meth-
           ods* 75.4 (2012), pp. 2209–2235.

[Neu65]   S. Neumark. *Solution of cubic and quartic equations.* English.
           (The Commonwealth and International Library) Oxford - London
           - Edinburgh - New York - Paris - Frankfurt: Pergamon Press.,
           1965.

[NH09]   Mitsuhiro T. Nakao and Kouji Hashimoto. "A numerical verifi-
           cation method for solutions of nonlinear parabolic problems." In:
           J. Math-for-Ind. (2009).

[NKK12]   Mitsuhiro T. Nakao, Takehiko Kinoshita, and Takuma Kimura.
           "On a posteriori estimates of inverse operators for linear parabolic
           initial-boundary value problems." In: *Computing* 94.2-4 (2012),
           pp. 151–162.

[Plu08]   Michael Plum. "Existence and multiplicity proofs for semilinear
           elliptic boundary value problems by computer assistance." In:
           Jahresber. Dtsch. Math.-Ver. 110.1 (2008), pp. 19–54.

[RLH00]   M. Raible, S. J. Linz, and P. Hänggi. "Amorphous thin film
           growth: Minimal deposition equation". In: *Phys. Rev. E* 62
           (2000), pp. 1691–1694.

[RMS13]   James C. Robinson, Pedro Marín-Rubio, and Witold Sadowski.
          "Solutions of the 3D Navier–Stokes equations for initial data in
          $\dot{H}^{\frac{1}{2}}$: robustness of regularity and numerical verification of regu-
          larity for bounded sets of initial data in $\dot{H}^1$ ". In: *J. Math. Anal.
          Appl.* 400.1 (2013), pp. 76–85.

[RS08]   James C. Robinson and Witold Sadowski. "Numerical verification
          of regularity in the three-dimensional Navier-Stokes equations for
          bounded sets of initial data." In: *Asymptotic Anal.* 59.1-2 (2008),
          pp. 39–50.

[SP94]   M. Siegert and M. Plischke. "Solid-on-solid models of molecular-
          beam epitaxy". In: *Physical Review E* 50 (1994), pp. 917–931.

[Thu94]   William P. Thurston. "On proof and progress in mathematics".
          In: *Bull. Amer. Math. Soc. (N.S.)* 30.2 (1994), pp. 161–177. ISSN:
          0273-0979.

[Win11]   Michael Winkler. "Global solutions in higher dimensions to a
          fourth order parabolic equation modeling epitaxial thin film
          growth". In: *Zeitschrift für angewandte Mathematik und Physik*
          62.4 (2011), pp. 575–608.

[Zgl10]   Piotr Zgliczyński. "Rigorous numerics for dissipative PDEs. III:
          An effective algorithm for rigorous integration of dissipative
          PDEs." In: *Topol. Methods Nonlinear Anal.* 36.2 (2010), pp. 197–
          262.

In der Reihe *Augsburger Schriften zur Mathematik, Physik und Informatik,*
herausgegeben von Prof. Dr. B. Aulbach, Prof. Dr. F. Pukelsheim,
Prof. Dr. W. Reif, Prof. Dr. B. Schmidt, Prof. Dr. D. Vollhardt,

sind bisher erschienen:

Alle erschienenen Bücher können unter der angegebenen ISBN im Buchhandel oder direkt beim Logos Verlag Berlin (www.logos-verlag.de, Fax: 030 - 42 85 10 92) bestellt werden.